Web Watching

A GUIDE TO WEBS & THE SPIDERS THAT MAKE THEM

Larry Weber

Stone Ridge Press
2515 Garthus Road
Wrenshall, MN 55797
www.stoneridgepress.com
thesparkygroup@gmail.com

WEB WATCHING

A Guide to Webs & the Spiders that Make Them

Printed in Manitoba, Canada by Friesens
10 9 8 7 6 5 4 3 2 1 First Edition

ISBN-13: 978-0-9909158-7-4

Graphic Designer: Mark Sparky Stensaas of Stone Ridge Press
Illustrations by Rick Kollath [www.kollathdesign.com]
& Billy Anderson [www.billyandersonart.com]
Cover photos by Larry Weber & Sparky Stensaas

Suggested Retail $16.95

Table of Contents

Acknowledgements

Sorting through and helping to compile my thousands of photos of spider webs was a massive undertaking. The project was made easier with help from Mark Sparky Stensaas. He listened to me when I told him of my plans and gave support on how to put out a book about Web Watching. He was able get some needed photos for me and solved many of my computer problems. Thanks to Rick Kollath for the use of his amazingly detailed illustrations. And to Billy Anderson of the Art Farm who recreated some illustrations for me. Chad Heins and Cassie Novak also contributed photos and knowledge for this book. John Geissler let me do web watching and provided the orb webs sign on the last photo in the book. —Larry Weber (October 2017)

Dedication

To my companions Frannie and Isaiah who often were watching me as I was doing my Web Watching.

And to...

John Geissler who was my partner on many classes involving Spiders and Web Watching.

A Guide to Web Watching

Nearly all of us have had an experience such as this...We are out walking, biking, jogging or commuting to work on a summer or fall morning and as we move towards the rising sunlight, we see dew-covered webs along our route. Dew drapes the grasses, flowers and shrubs here and among these droplet-coated plants are a plethora of webs constructed at these sites.

If we look more than with just a passing glance, we'll note just how many there are, and that they are very diverse. The webs are not all the same size; they also vary in their shapes and locations. Most of us would recognize that they are made by spiders, but we usually don't go much beyond that. And though the webs are easy to see, their makers are not. They seem to have emerged from nowhere. Who made them? When? Why? Where are the spiders now? Many people do not appreciate the spiders themselves, but nevertheless see these webs as beautiful. Webs are highly photogenic and their appearance is often seen as being synonymous with late summer or early fall mornings.

Spiders are not the best-loved critters that live with us, even though they are very common and we often are living near them. Spiders are frequently associated with webs. Indeed, the word "spider" is a derivative of the Dutch word "*spinder*" in reference to their spinning abilities. Many spiders do make webs, but more spiders do not. However, all spiders have the silk glands and spinnerets needed to make and use silk.

Spiders are not the only web makers amongst us, though they are most likely the species we're seeing. Other organisms that make webs from threads are types of insects, especially caterpillars. Ones that we are likely to see in traveling near our residents are those made by Fall Webworms and Eastern Tent Caterpillars (see photos next page). While Fall Webworms construct theirs on the outer branches of small trees, the Tent Caterpillar webs stay close to the main part of the plant. And, of course, moths are well known makers of silken cocoons.

In this book, we take a closer look at arachnid webs and the spiders that make them. When examining the webs more closely, we can learn to recognize the spiders by their webs. (It is often hard to see the spider itself, even if it is in or near the web.) This could be compared to determining kinds of birds or insects

Not all "webs" are made by spiders; The tent caterpillar (*Malacosoma americanum*) makes this "web" (left) and the Fall Webworm (*Hyphantria cunea*) creates this web (right).

by their calls and songs without actually seeing the critters. Or maybe it might be likened to looking at animal tracks and recognizing who has passed this way.

Webs can be found in our yards, woods, roadsides, fields, swamps and bogs. They are most abundant in late summer, but are present from spring until late fall. I have already seen webs in April and as late as early November here in northern Minnesota. Mostly we see the snares used to catch food; usually insects, but they are not the only constructions made by spider silk. All spiders, not just the web builders, have spinnerets; the organs used for web making. Internal silk glands produce different types of silk that are utilized throughout the year. The webs we see are only part of the spider story. In order to take a complete look at the webs and silk thread used, we need to take a closer look at the spiders themselves.

Spider Classification

Like other kinds of organisms, spiders have been classified into groups that further define them. All animals are put into large groups called phyla based on scientists' beliefs about their genealogical (phylogenetic) and physical features that relate them to each other. Spiders belong to the largest phylum; the arthropods. Members of this phylum have segmented appendages and a hard outer skeleton (***exoskeleton***). This incredibly huge phylum includes four familiar groups called classes; crustaceans (crabs, shrimp, crayfish, water fleas, etc.), myriapods (centipedes and millipedes), insects (flies, butterflies, bees, dragonflies, beetles, etc.) and arachnids (spiders, scorpions, ticks, etc.). Arachnids all possess eight legs.

In addition to spiders, ticks, mites, pseudoscorpions and daddy-long-legs (har-

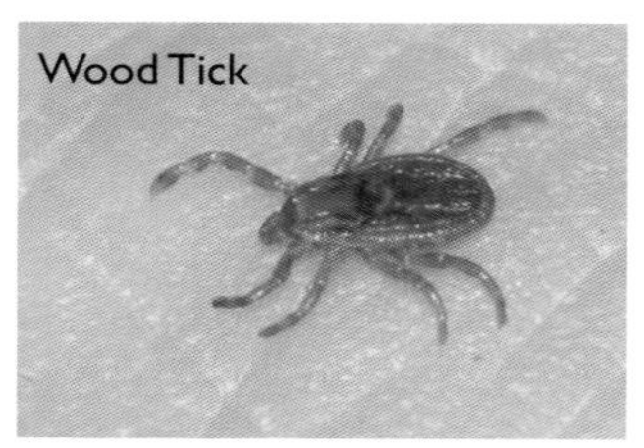

Here are some other of our northern arachnids; all have 8 legs like spiders.

vestmen) are some Arachnids that we may find in our neighborhood. Spiders differ from these other arachnids in various body features, even though all have the diagnostic eight legs. Within class Arachnida, all spiders belong the an order called Araneae.

Within the order Araneae, spiders are further divided into two infraorders; mygalomorphs (orthognatha) and araneomorphs (labidognatha). Basically, the difference is how they use their fangs. The larger mygalomorphs injects them downward into the prey. Included in the group are tarantulas and trap-door spiders. By strict definitions and body features, they are not true spiders. There are no members of this infraorder in our region. The smaller araneomorphs clasp prey with fangs coming from the sides; a horizontal motion. Our local spider species are all araneomorphs; true spiders.

Spider and Insect Comparison

Insects superficially resemble spiders, but a closer look reveals many differences. Insects have three body parts; head, thorax and abdomen. Spiders have only two body parts. The head and the thorax are fused to form the ***cephalothorax***, which is connected to the abdomen. Insects have three pairs of legs and most have two pairs of wings; all attached to the thorax. Spiders have no wings and four pairs of legs which are all attached to the cephalothorax.

Insects have antennae while spiders do not. Spiders; however, have another pair of appendages, the ***pedipalps***, that are attached near the head. Insects have large compound eyes composed of many small lenses plus three tiny simple eyes while the typical spider has eight eyes (rarely six). Insect mouthparts are varied

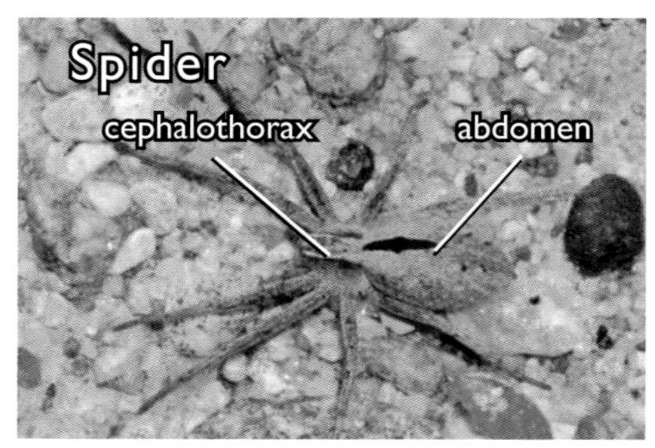

Spiders such as the Diamond Spider (*Thanatus vulgaris*) have eight legs and two body parts, while insects (like this ground beetle on right) have six legs and three body parts.

and complex and different from those of spiders. While many insects have biting or piercing mouthparts, they do not have ***fangs*** as found on the jaws (***chelicerae***) of spiders. These fangs are connected internally to ***venom glands***; not found in insects. Moving to the hind end of the body, many insects have stingers and/or ovipositors for laying eggs, but lack the spinnerets and the internal silk glands as seen in spiders. It is this apparatus that allows the spiders to make and use webs.

More differences between spiders and insects are in their life cycles. Insects go through a physically radical metamorphosis in which the young (larvae) change into adults that are frequently completely different. Many also go through a non-motile pupal stage. Immature spiders go through several ***molts*** until becoming adults, but the young look very much like miniature adults.

While many insects feed as predators, many others are herbivores. Spiders are carnivores. Their animal prey may be quite diverse but they are not plant feeders. Many spiders use their silk material to construct snares to catch insects. And though we often associate spiders with webs; a large number of kinds do not use this method to catch their prey.

Spider External Anatomy

Most of us are quick to recognize a spider when we see the specific features of eight legs and two body parts. Though frequently drawn incorrectly in cartoons, spiders have all eight legs attached to the front of these two body parts, the ***cephalothorax***. The rear part of the body, the ***abdomen,*** is usually the larger of the two parts, but has no legs attached.

The Cephalothorax

The cephalothorax is composed of a head (*cephalo*) and a thoracic region (*thorax*). When viewed from above, the top, or dorsal, portion is known as the ***carapace***. The underside, or ventral, is called the ***sternum***. Usually, we see the spider's carapace, but with many web dwellers, the sternum can also be seen since placement in a web is often inverted. Though the carapace is often without patterns of colors, many spiders have a central lower spot; the ***thoracic furrow***. This longitudinal depressed area is where the internal muscles of the ***sucking***

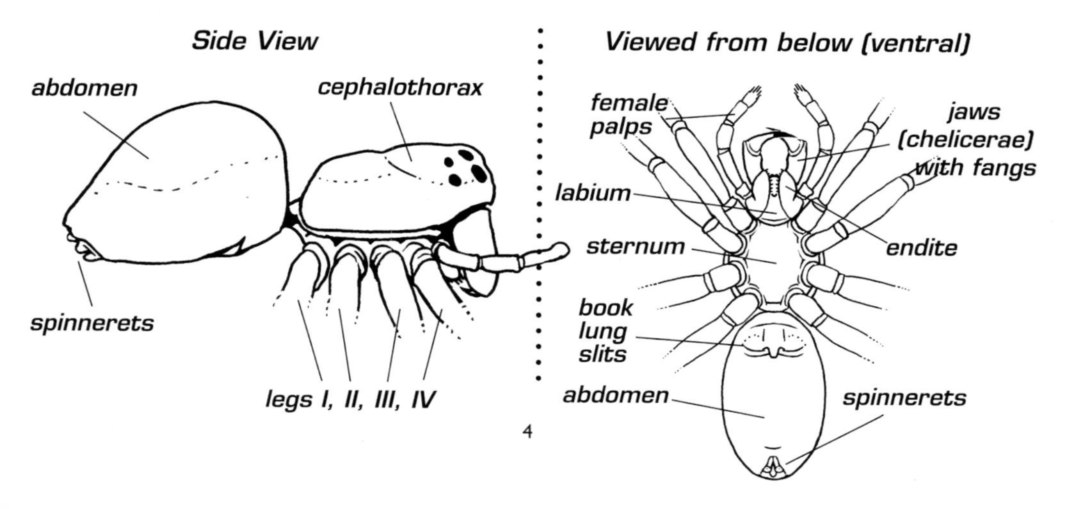

stomach are attached. Such powerful muscles are needed for its feeding method.

The cephalothorax has more than just legs attached to it. The ***head***, as seen from the front, has the same name as we use for the front of our head; the ***face***. There are two main facial features; an ***ocular region***, the area around the eyes, and the ***chelicerae***, often called ***jaws***, below the eyes. The vertical chelicerae are usually lined with pointed structures, called teeth, and ***fangs*** are at the tip. ***Venom*** is secreted from the fangs into the spider's prey.

Slightly back and down from the face is a pair of appendages; the ***pedipalps*** (often shortened to "palps"). In females the palps are merely segmented organs that look like small legs. But in males, the last segment is swollen and modified to serve as a semen receptacle, used in mating. They are fully developed only in mature males. Held out like little "boxing gloves," the male spiders are easy to discern from the females with a little practice. Males are usually smaller than females and the ones that we are likely to see in or near webs are the females.

Ventrally (underneath), in front of the sternum is the spider's mouth. It consists of a simple opening that leads to the ***pharynx*** and ***stomach***. Externally, the mouth has flattened structures called ***endites*** on each side and a ***labium*** (lip) behind.

Spiders usually have eight eyes. Though this is the normal situation, eye numbers in species around the world can vary from eight to six to four to two and even some with none. Nearly all the ones in our region have eight eyes. These eyes are

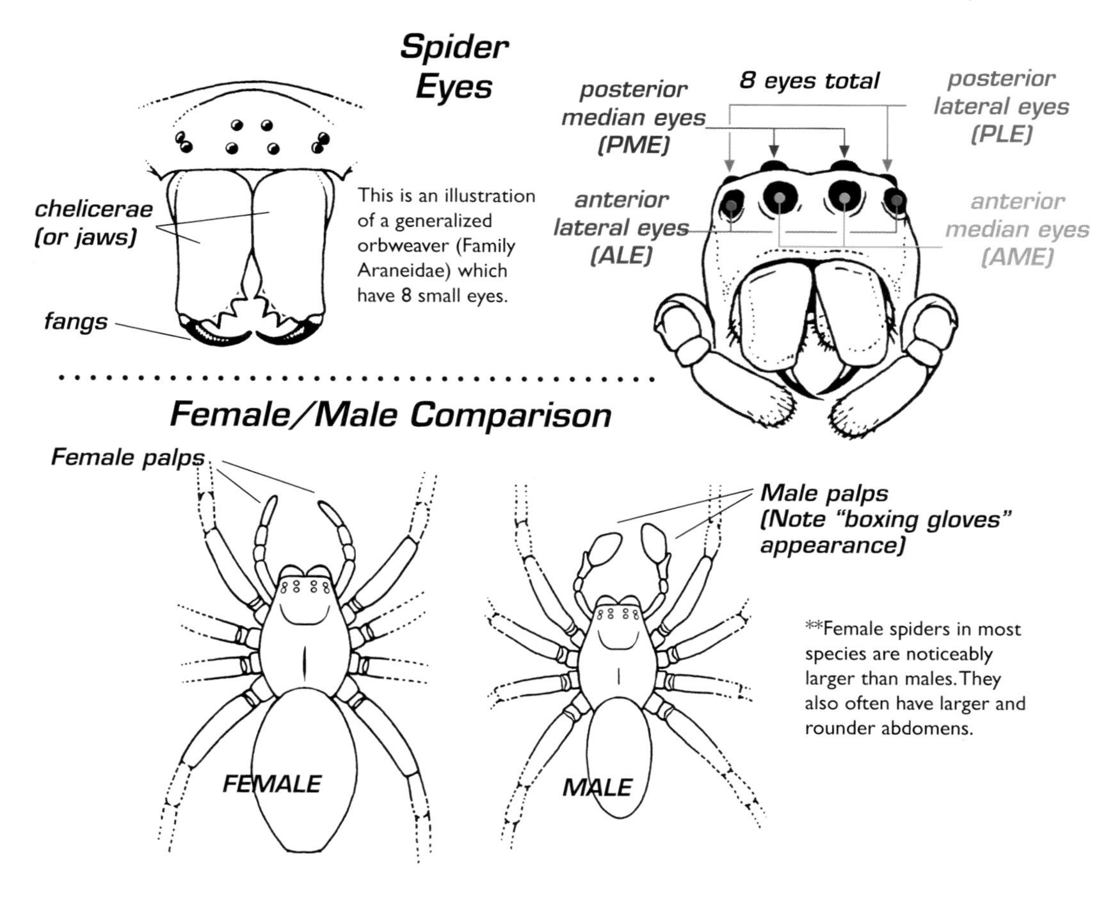

normally arranged in two rows; the anterior, towards the front and below the ocular region. The posterior row is above or behind these. Eyes are grouped in pairs and named for their location. Those in the middle of the front row are called the ***anterior median eyes*** (AME) and those to their sides are the ***anterior lateral eyes*** (ALE). Similarly, among the posterior eyes are the ***posterior median eyes*** (PME) and the ***posterior lateral eyes*** (PLE).

In many spiders, all eight eyes are similar in size, shape and function; but exceptions are common. Often the rows of eyes are not straight, but curved either towards (procurved) or away from (recurved) the front.

Basically, spiders are either sedentary hunters; often in a web, or active hunters that do not construct or use webs. Nearly all web hunters have poor eyesight, but active hunters often have large eyes with good sight. Most noticeable of the latter are wolf spiders (Lycosidae) and jumping spiders (Salticidae). While wolf spiders have large posterior median eyes (PME), the jumpers have big anterior median eyes (AME). Wolf spiders are usually nocturnal hunters while the jumping spiders pounce on their prey in the daytime. This behavior and their mating habits seem to indicate that jumping spiders have excellent sight and can see color. Spider eyes and their arrangement vary enough among species that they can serve as a help in identification. As part of the face, they help in spider recognition.

For many people, no part of the spider's anatomy is more noticeable than the legs. Because of our stereotypes, we may assume that all spiders have very long and hairy legs. But a closer look will show us that spider legs range from short to long, thick to thin, hairy to smooth.

Traditionally, spider length has been measured from the face to the spinnerets on the rear; basically body length excluding the legs. Though this is a useful measurement, body length is not what most people think of when describing a spider. In this book, we also use the leg span to describe overall spider size. It is measured as the distance from tip of the front legs to the tip of the hind legs.

Legs are consistently composed of seven segments, more than most other arthropods. Starting from the cephalothorax and going out, the parts are called ***coxa, trochanter, femur, patella, tibia, metatarsus*** and ***tarsus***. With the exception of the first two, the names are similar to those of human leg anatomy. At the far end of the tarsus, spiders possess ***claws***. Depending on the species, there are either two or three claws. Web-making spiders and a few wandering hunters have three claws. The two-clawed spiders do not make webs, but can be either active or sedentary in their hunting.

Spider legs are usually covered with hairs, but these hairs can show quite a bit of structural variation. Most numerous are the simple, short hairs that serve a tactile function. Scattered among these hairs may be short sharp spines and /or long thin hairs called ***trichobothria*** which are especially sensitive to surrounding conditions. This tarsal hairs called ***scopula*** show up on some species. Members of one family, the cobweb weavers (Theridiidae), have thick bristles on the tarsus. Used to pull out silk to wrap over struggling prey, these "combs" lead to the other common name for this family; the comb-footed spiders. Many spiders also have tufts of hair among the claws at the end of the tarsus.

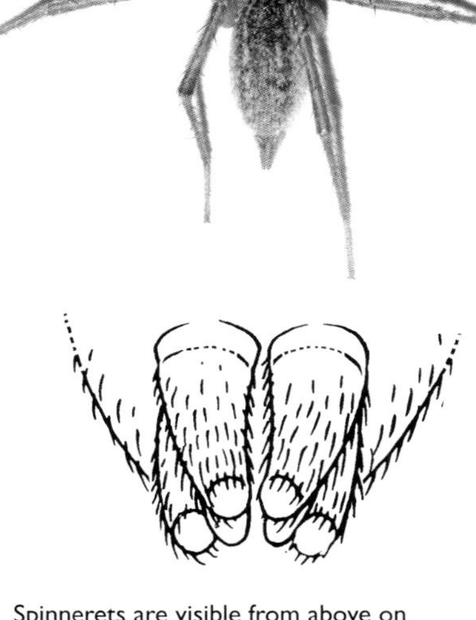

Spinnerets are visible from above on a few spiders such as the *Agelenopsis* grass spider (top photo). Illustration of spinnerets from below.

The Abdomen

The abdomen is attached to the cephalothorax by a small tube called the ***pedicel***. Usually much larger than the cephalothorax, the abdomen ranges from being just a long thin organ to being big and ball-shaped. Typically the abdomen is oval when seen from above. This is the part of the body most noticed in the field, and fortunately, it is often marked with spots, bands, chevrons, stripes or other patterns that help us identify the spider to genus or species.

Though no legs attach to the abdomen, it does have its own appendages. Connected to the undersides of the terminal end, but sometimes still visible from above, are the ***spinnerets***. We can usually only see one pair in the field, but there are actually three pairs. Normally, the first and third pairs are the longest with the second pair being very small. These structures are unique to spiders and as the name implies, they function with spinning webs and other uses of silk. Agile and flexible like fingers; spinnerets are used in ejecting and placing silk from the internal silk glands.

One unique group of spiders has another abdominal structure near the spinnerets called the ***cribellum***. The cribellum spins a special silk called "hackled band threads." Such spiders also heave a thick line of hairs, the ***calamistrum***, on the metatarsus. The cribellate spider families in our region include Uloboridae (cribellate orbweavers), Amaurobidae (hackledmesh weavers) and Dictynidae (meshweb weavers).

Also on the underside are several orifices of importance. Two transverse lines in the fore region are the slits that open to the internal breathing strata known as ***book lungs*** (spiders also have ***spiracles***). Also nearby are the genital openings. In females, this opening is often covered or partially covered by a growth called the ***epigynum***. It is species specific and connects with the male's pedipalp tip, or ***bulb***, in mating. The different epigynum forms are used by experts to determine spider species. Such an undertaking is difficult and demands the use of a microscope. Two other tiny openings are here too: The ***anus*** from the digestive system and the ***spiracles*** that lead to trachea and aid in respiration.

Some of the larger spiders appear to have indented depression on the upper surface of their abdomen. These dimples are actually muscle attachment points from the inside of the exoskeleton.

Spider Internal Anatomy

To most observers, the spider's internal anatomy is a mystery, but a look at the parts and their function can help us better understand their behavior.

The nervous system, including brain and optic nerves, stomach and venom glands take up most of the inner cephalothorax. Of special note in the cephalothorax are the sucking stomach and the venom glands. After catching prey, the spider uses its fangs to inject venom flowing through a tube from the venom glands to the end of the fangs. After the prey is subdued, enzymes are further injected from the mouth into the prey insides. These powerful chemical serve to liquefy the organs. The spider now has an "insect milk shake" within the prey's exoskeleton. They then suck up their meal through the mouth, through the pharynx and into the stomach. The powerful stomach muscles required to do this are attached to the insides of the carapace; the site called the thoracic furrow (which is visible on the carapace). Larger spiders will use their legs and feet to crush the exoskeleton of their prey as they feed, while smaller ones leave the outer shell intact.

With the exception of the family Uloboridae (cribellate orbweavers), all spiders in our region possess ***venom glands.*** Tubes that open at a pore near the tip of the cheliceral fangs reach down from the glands. Cylindrically shaped, these

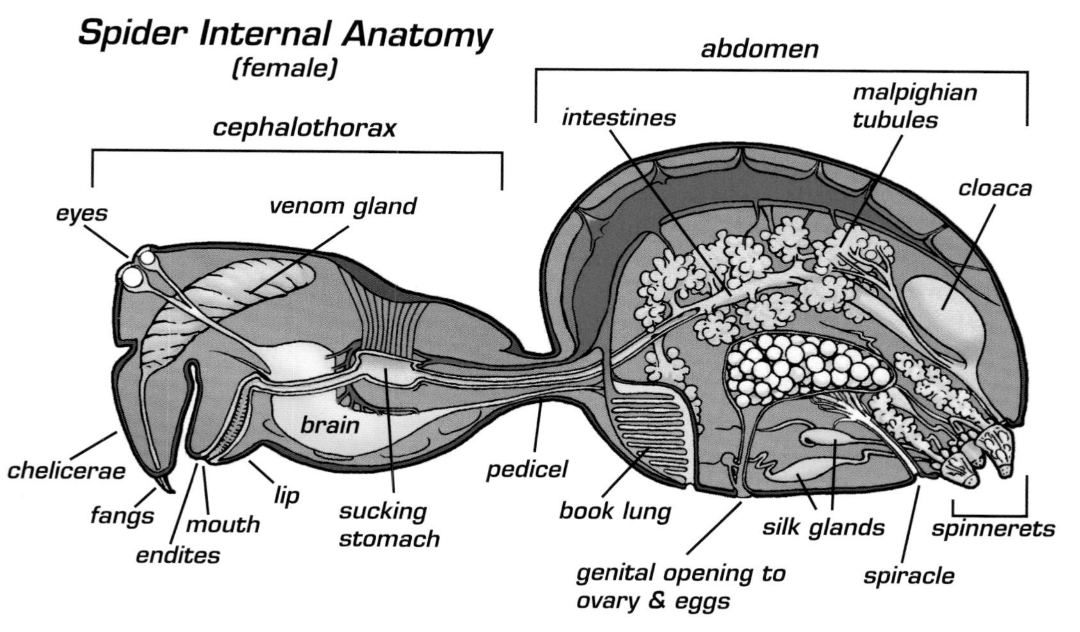

venom glands fill the internal space above the mouth, behind and below the eyes. In some species these glands are so small that they hardly go beyond the base of the chelicerae. Venom produced by a spider may vary with their age and physical condition. Only a few species have venom virulent enough to be harmful to humans, and none regularly live in our region. Remember, spiders are timid and even if provoked to attack, most have fangs that are too tiny to puncture our skin.

Inside the abdomen of a female spider (what we are most likely to see since the males are small and not usually even in webs), we find the heart, midgut (intestines), ***malpighian tubules*** (primitive kidneys), ovaries, respiratory organs (book lungs and trachea) and the ***silk glands*** with tubes leading to the spinnerets. The numbers and kinds of silk glands may vary but normally from five to seven kinds. Since the abdomen contains so many vital organs, any damage to it can lead to a quick death for the spider.

Spinnerets and Silk

Located on the under or ventral side of the abdomen, near the rear end, are six small appendages; the ***spinnerets***. These structures are unique to spiders and they are present on all kinds of spiders; even those that do not construct webs.

When first seen on the underside of the abdomen of a spider, the spinnerets look like just a pair of caudal organs. For most spiders, they cannot be seen when viewed from above. (The most obvious exception to this is the funnel weavers; the Agelenidae, where they are easily seen and appear to be two "tails.") A closer look reveals three pairs of spinnerets. Located where they are, they are named anterior (towards the front), median (middle) and posterior (furthest back). Typically, the median are smallest of the three. Largest is the anterior, composed of three segments; posterior is two and the median is only a single segment. Equipped with many muscles, the spinnerets are very mobile. They also have tiny openings called spigots on the surface. These spigots are the terminal ends of the various silk gland tubes from inside the abdomen.

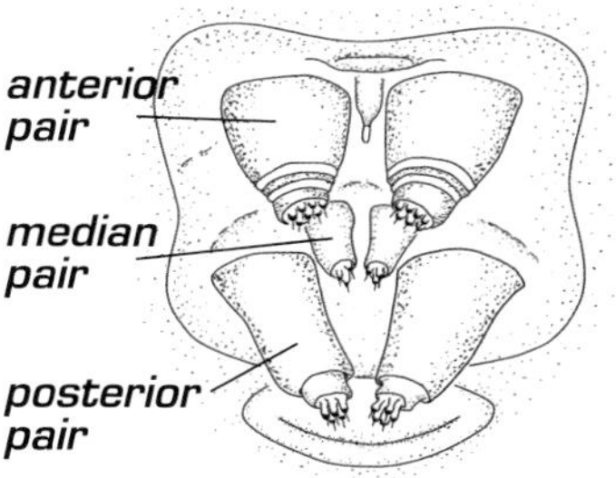

Spinnerets shown from dorsal posterior view. Most spiders have six spinnerets in three pairs.

Illustration on the opposite page shows the internal anatomy of a female spider. Spinnerets are external, but are attached to the tubes coming from the internal silk glands. These are not a single type of glands, but several types with each kind of spider. Spiders have five to seven types of silk glands in the abdomen. We will take a closer look at the seven kinds of glands as found in orbweavers.

The glands are frequently named after their shapes. Each is attached to spinnerets. Some silk glands have tubes that go to two different spinnerets.

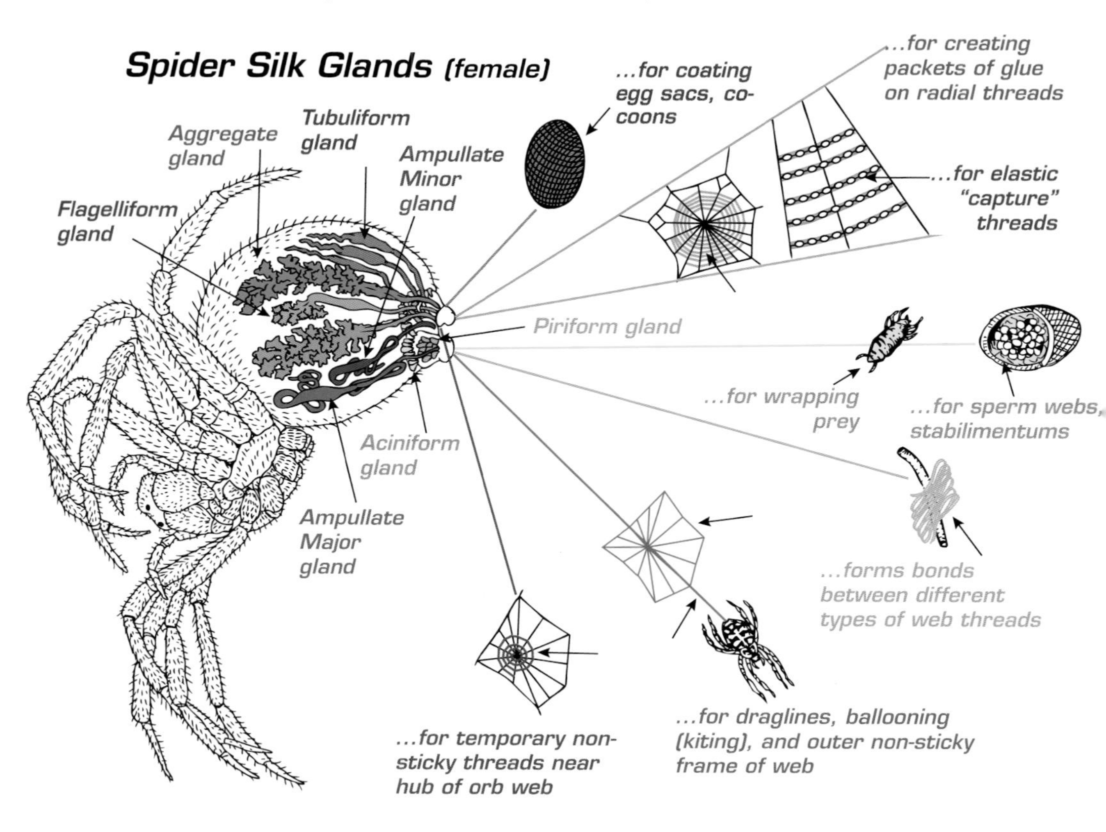

Depending on the species, spiders have five to seven silk glands. Above is a generalized color-coded illustration of these glands and their typical uses. See text on pages 9-13 for details on what each gland produces.

Attached to the Anterior Spinnerets are these silk glands:

- Ampullate Major [draglines; ballooning or kiting; outer non-sticky rim of web]
- Ampullate Minor [temporary non-sticky threads near hub of orb web]
- Piriform (Pyriform) [forms bonds between different types of threads in web]

Attached to the Median Spinnerets are these silk glands:

- Ampullate Major [draglines; ballooning or kiting; outer non-sticky rim of web]
- Ampullate Minor [temporary non-sticky threads near hub of orb web]
- Tubuliform (cylindriform) [coating of spider egg sacs; cocoon construction]
- Aciniform [wrapping prey]

Attached to the Posterior and longest Spinnerets are these glands:

- Flagelliform [elastic "capture" spiral threads of web]
- Tubuliform (cylindriform) [coating of spider egg sacs; cocoon construction]
- Aciniform [silk to wrap and secure prey; sperm webs; stabilimentums]
- Aggregate [makes "packets of glue" to make radial capture threads sticky]

These seven different silk glands produce various kinds of silk; having different functions. When seeing all the uses of silk, we get a better appreciation and understanding of the uniqueness of this spinneret-silk gland apparatus present in all species of spiders.

- Silk from the Ampullate Major gland is used by spiders as they lay down a dragline while moving; usually walking. The same threads are what creates the gossamer threads we see as they disperse in what is called ballooning (maybe more accurately known as kiting). Besides this aid to spider movements; orbweavers use this silk to lay down the outer rim of their webs along with the radii (non-sticky) threads.
- Silk from the Ampullate Minor gland makes threads used in orb web construction. Near the hub, in the center, the spider attaches this non-sticky thread to radii to make a temporary scaffolding. It is later taken down when the spiral threads are attached to the radii.
- Silk from the Flagelliform glands produce the captive spiral silk threads. It is this spiral that is layed down on the web from the outside towards the center that actually captures the prey. Though these threads are themselves non-sticky they do hold "packets of glue," from aggregate glands. They are highly elastic.
- Silk from the Aggregate gland is the material that makes the spiral threads of an orb web sticky. Spiders deposit drops of glue from these glands on the spirals.
- Silk from the Piriform gland is used to form bonds between separate threads in the web; the radii threads attached to the spirals; making the web more stable.
- Silk from the Aciniform gland is used to make threads that are used to wrap and secure the freshly caught prey (usually insects). This strong thread is also

Argiope wrapping its prey with silk. This type of silk is from the Aciniform gland. It is strong and perfect for subduing prey.

used in the stabilimentum that are present in the webs of some spiders. And it is used by males to form their sperm webs.
- Silk from the Tubuliform gland is used in the coating of spider egg sacs. Some spiders also construct a cocoon for themselves and eggs or as a hiding place.

Spider Silk

Spider silk is a most remarkable substance. Formed as a complex protein liquid within the glands, it quickly becomes an elastic thread when outside the body. Not only does silk have a tensile strength (the amount of force required to tear a material) greater than tendons per weight, but it is also so elastic that wet threads can stretch to nearly twice its length before breaking. The tenacity of spider silk is slightly less than that of nylon yet is twice as elastic. Silk proteins are largely eaten by orbweavers from their old webs and reused to make new silk threads the following day. Radioactive tagging has indicated that up to 70 percent of the initial web material shows up in the new web, even though it may be only a half hour between destruction and new construction. Apparently, the chemistry of the silk allows for quick digestion and rebuilding.

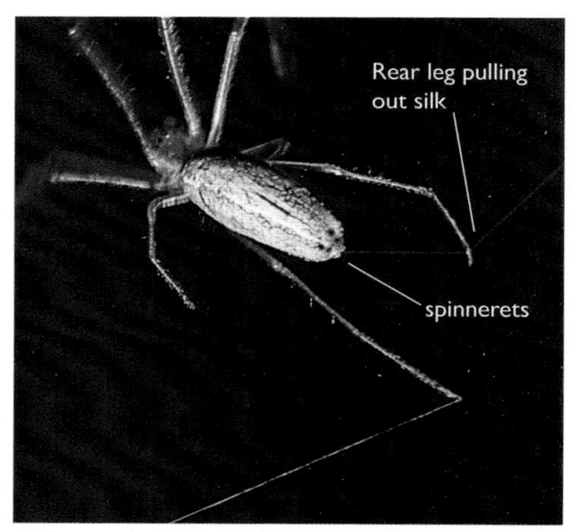

A *Tetragnatha* spider in the midst of spinning an orb web. Note how the hind legs help in pulling out the silk from the spinnerets.

Though largely proteins, other compounds are found in the spider silk as well. These include sugars, lipids and pigments that might affect the aggregation behavior and act as a protection layer in the final fibers.

The production of spider silk differs in an important respect from the production of most other fibrous biological materials. Rather than being continuously grown as keratin in hair or cellulose in plant cells, it is "spun" on demand from liquid silk precursor sometimes referred to as unspun silk dope; from specialized glands.

The spinning process occurs when a fiber is pulled away from the body of a spider, be that by the spider's legs, by the spider's falling and using its own weight, or by any other method including being pulled by humans. The word "spinning" is misleading as no rotation of any component occurs, but the word comes from when it was thought that spiders produced their thread in a similar manner to the spinning wheels of old. In fact the process is more accurately a pultrusion; similar to extrusion, but the force is induced by pulling at the finished fiber rather than being squeezed out of a reservoir of some kind. Most spiders pull the threads from the spinnerets with their fourth pair of legs. The unspun silk dope is pulled through silk glands, of which there may be both numerous duplicates and also different types on any one spider.

An example of the complexity and coordinated effort needed by a spider can be seen by looking at the spinneret-silk gland apparatus of an adult Cross Orbweaver (*Araneus diadematus*). It consists of the silk from the following glands:

- 500 silks from Piriformes for attachment points
- 4 silks from Ampullaceae for the web frame
- About 300 silks from Aciniformes for the outer lining of egg sacs and ensnaring (wrapping) prey
- 4 silks from Tubuliformes for egg sac silk
- 4 silks from Aggregate for glue
- 2 silks from Flagelliformes for the thread of glue lines

Types of Webs

Webs of all types are made for the purpose of catching prey. The prey is usually insects. This six-legged food source is so abundant that there is always much to be found. Some insects are caught in the daytime (e.g. grasshoppers); others at night (e.g. moths).

In order to be a successful snare to catch food, the spider webs need to have a few adaptations. The webs need to be strong enough to hold the struggling insects. The threads must be thin while still being strong so as not to be seen easily and thus avoided by insects. In addition to this, many spider webs are also equipped with sticky glue-like material that helps subdue the captured insect. And since the snare is constructed with the intent of catching food, the web must also have a site where the spider may remain to see, or, more likely, feel the prey. Spiders need to be able to move on their own web; hide in a shelter nearby and come onto the web to wrap up the prey without getting caught in their own web. And often they take down the web; eating the threads. The remarkable spider silks of several kinds allow all this to happen. Spiders have evolved to make snares that are of different types.

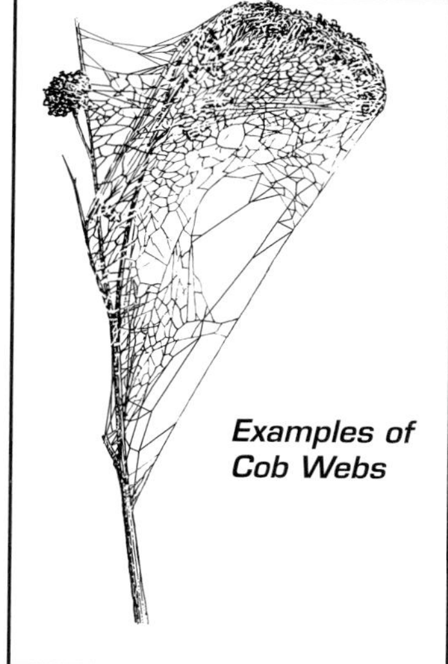

Examples of Cob Webs

Webs are often what we notice in our acquaintance with these eight-legged critters. Spiders are predators and the constructed webs that we are likely to see are formed as ways of catching their prey. There are variations in the forms of webs, but only four basic types: cob webs, sheet webs, funnel webs and orb webs.

Cob Webs

Cob webs, sometimes called irregular webs, are the ones that we frequently see indoors; perhaps in the corner of a room, by a window sill or heat duct. They are also quite common outdoors as well; near buildings, but often in plants too. Though they look like a confused mess of threads, they are constructed by the spider as a way of catching food as well as providing a place of safety for the web maker. The jumble of threads slows and contains the insect enough for the spider to make its catch. Lower threads often have some stickiness as well.

In the North Woods, we have three families of spiders that use cob webs to catch meals: Theridiidae (cobweb weavers), Pholcidae (cellar spiders) and Dictynidae (meshweb weavers).

Theridiidae are the most common cobweb spider. Webs are usually indoors. We often see them in the house, garage, basements and other outbuildings. I have also seen them on the outsides of buildings and in trees and other plants. Spiders hang inverted in the webs. The large number of threads looks like a mess, but the owner finds its way around. Some of the lower threads are often sticky.

Pholcidae (cellar spiders) have adapted to life indoors so well that their webs are nearly always inside. Less commonly webs may be outdoors—near buildings, in caves, on cliffs. They remain active all year. Spiders hang inverted in the web and when disturbed move their bodies so fast as to give the appearance of just a blur.

Dictynidae spiders are a family not so well known, but in late summer, they are extremely common in old plants in fields. Webs are most likely on the top of dead stalks standing in open lands. I find many in late summer but few at other times of the year. Spiders that make these webs are very small and hide in the web. With a little searching, they can be located. I have not seen them indoors.

Sheet Webs

Sheet webs are constructed in different shapes and are very common and photogenic when seen on a dewy morning. I have not found these webs indoors. They are mostly the snares of one family: Lyniphiidae (sheet-web weavers). This is an extremely large family of very small spiders. Within this family are a few that are larger and they make the webs that we associate with this group.

Examples of Sheet Webs

Webs can vary in shape from those that look like a bowl with a flat section beneath (bowl and doily; cup and saucer) to some that are inverted bowls (domes) while still others are more flat and are called hammock webs. Webs are common in shrubs (especially evergreens), yards, swamps, bogs or woods. I have also found them in meadows and fields, but not as commonly. Webs may last for many days or weeks. Spiders remain in the webs, inverted on the undersides of the web, be it "bowl," "dome," or "hammock."

Another family, Hahniidae (dwarf sheet spiders), is composed of tiny spiders that live on the ground. Some make small flat webs just a couple of inches in diameter. I also call these sheet webs.

Examples of Funnel Webs

Funnel Webs

Constructed much like that of a sheet web, funnel webs differ by having a hole either in the center or to one side. This opening is where the spider sits when waiting for prey. Backing into this opening, the spider can also take shelter from danger. The shape of this web, with a tubular hole in the center, is why it is called a funnel web.

Two families are regarded here as funnel web makers: Agelenidae and Amaurobiidae.

By far the best known of the funnel-web spiders are the Agelenidae (funnel weavers). They make webs in our lawns in summer and often later, as the weather gets colder, we might see them indoors. Webs may be up to one foot across with an obvious opening. Looking closely, we can often find the spider sitting here. Spiders are brown, often with stripes, and may be confused with wolf spiders, but have smaller eyes. They are one of the few spiders that we can see the spinnerets sticking out like a pair of tails when seen from above.

Examples of Orb Webs

Amaurobidae (hackledmesh weavers) are less known. They live under logs, boards and rocks. Though seldom seen, I find they are fairly common. Webs are much smaller than Agelenidae and of a mesh-type of threads. They have a centralized openings as well, and the dark spider rests and waits for prey here.

Orb Webs

When most of us think of spider webs, it is these circular orb webs that we have in mind. Not only are they common, they are easy to see; especially in late summer when they hold the morning dew drops on their threads. At this time, they also become highly photogenic and even people who do not appreciate spiders still find these intricate webs to be things of great beauty.

Most orb-web spiders in our region are in the family Araneidae (orbweavers). Spiders range from small to large and my construct webs in our yards, gardens, shrubs, trees, fields, meadows, buildings and other man-made structures. The fans of *Charlotte's Web* will note that she was in this family. Though this family is well represented by many kinds in this region; often large with huge webs, Araneidae is not the only family that makes orb webs. Webs are complex and made of many kinds of silk; some of which is sticky.

Different types of silk for different uses

Two other orbweavers are the Tetragnathidae (longjawed orbweavers) and Uloboridae (cribellate orbweavers). Tetragnathidae is the one we are most likely to see. Many are residents of the wetlands and with all the swamps, ponds and lakes in the North Woods, we will often see their webs. Webs are constructed along the shores of these aquatic sites; on grasses, sedges, cattails and rushes. But I have seen many of their webs on dead shoreline trees, branches and docks. Besides thriving in wetlands, these spiders also make webs among the late summer plants of fields and meadows. Differing from the other orbweavers, their webs are often horizontal instead of vertical as in most Araneidae webs.

Uloboridae are small spiders that also make mostly horizontal webs. Unlike the spirals of the rest of the orbweavers, their webs are not sticky. I don't find their orbs very often and when I do, it is always a great discovery.

Other Uses for Silk

Web spinning is what most of us associate with spiders and their silk. However, the construction of webs–snares to catch prey–is only one of many things that spiders use their silk for. In addition to webs; of which there are several types, spiders also use their silk for making and laying down a ***dragline*** as they walk, and similar threads are used to disperse the young in what is often called ***ballooning*** or ***kiting***. They also use threads for building ***egg sacs,*** which may be in the webs, in hidden sites or held by the spiders themselves. A similar strong silk material is used to wrap and subdue their prey in the web (***prey wrap***). Some spiders build "hiding places" that may be referred to as ***cocoons***. At the edge of many of the orb webs, the owner may construct a secluded structure called a ***retreat***. This might be a folded leaf or a piece of bark or it may be a chamber made of the silk threads.

All of these need and use silk. One group of spiders, the ***nursery-web*** spiders, make rather elaborate coverings of silken thread to hide their eggs and later, the young, while the adults stand by to guard. In mating, male spider form a ***sperm web*** where the place semen to be taken into their pedipalps later. The web itself is produced from a variety of threads; some sticky, some not. And when all is done, spiders can and do use web threads for food.

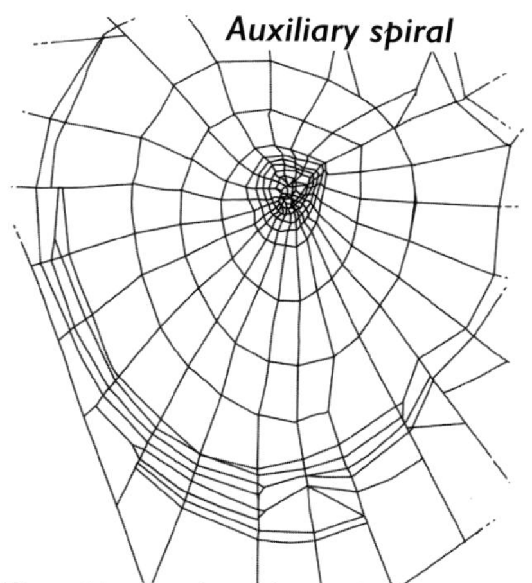

The spider lays down the auxiliary spiral first; it ties the radii together and helps stabilize the half-finished orb web.

Construction of the Orb Web

Those of us who have looked closely at a newly formed orb web are amazed at its intricacies. The various threads, elasticity and strength are superb. And even though we may not be seeing

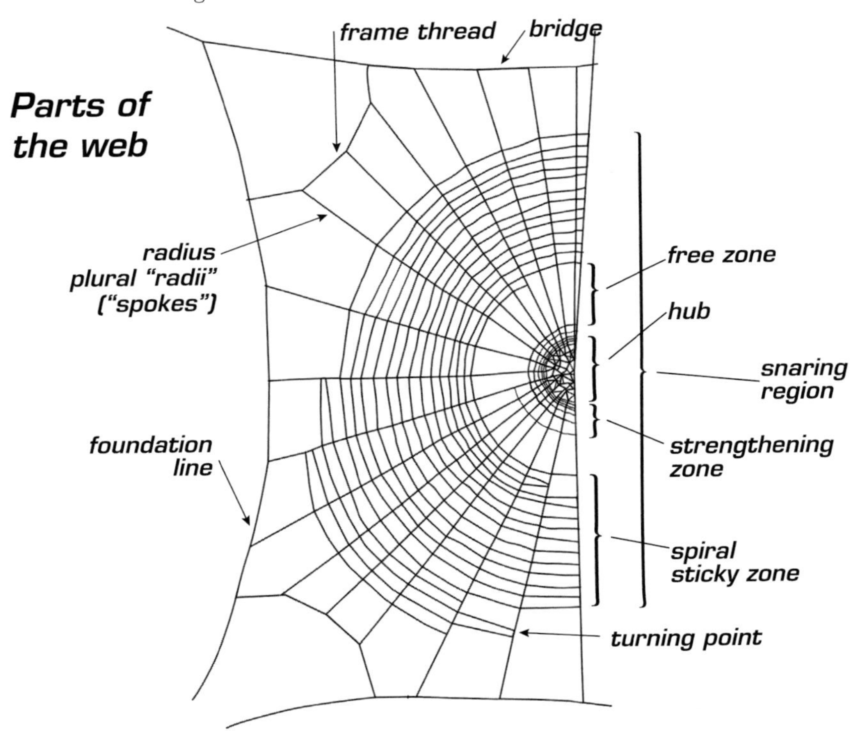

the formation process as we look at the final product, the orb becomes even more amazing when we note that these webs use silk of five or six different kinds (from five or six different silk glands) and that the webs are frequently made each day; usually at dusk, and can be constructed entirely in one-half to one hour. To do this job so efficiently, the spider needs to follow a regular routine that is made up of three phases of laying down the silken threads:

1. Frame and Radial Threads
2. Auxiliary or Temporary Spiral
3. Catching Spiral

To best understand this, we need to see all of the parts of a completed orb web. A typical orb web is composed of these parts:

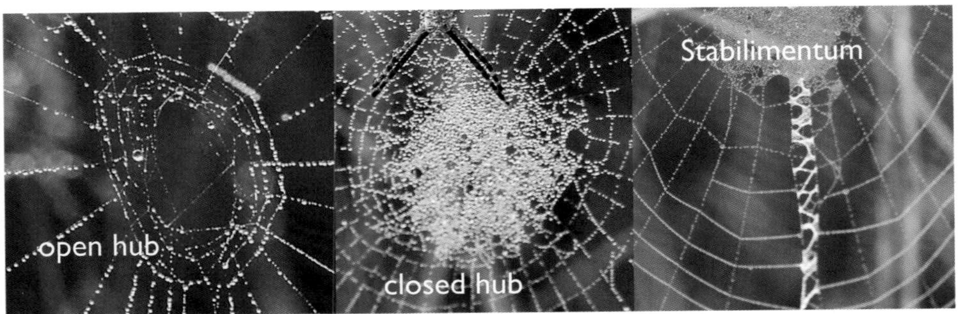

Different species of orbweavers create different styles of orb webs. Long-jawed Orbweavers in genus *Tetragnatha* make open hub webs (left), while the Banded Garden Spider (*Argiope trifasciata*) has a closed hub. Some *Argiope* also add a stabilimentum to the web.

- ***Outer Frame*** or ***Scaffolding***. These threads around the outside of the web are attached to a substrate. They are made from silk from the Ampullate Major gland.
- ***Radii*** (Occasionally they are called spokes). These are the threads that go from the outer peripheral of the web, attached to the frame, and go the center (hub). These threads are also made from silk of the Ampullate Major glands.
- ***Capture Spiral***; also called Sticky Spiral. These are the threads that are laid down in a spiral arrangement from the outer edge, attached to the radii and circle towards the center. Normally, they do not touch the hub. Made of silk from the Flagelliform glands, with added sticky globules from Aggregate glands. It is these viscid threads that actually catch and hold the prey. Typically, prey are caught in the stickiness and subdued here.
- ***The Hub.*** The center of the web, the hub, may consist of a series of non-sticky interconnected threads. Many spiders will sit on the hub in an inverted position while waiting for prey to hit the web. Some spiders construct a web in which threads in the hub are removed leaving a hole in the center; an open hub. I find the presence of an open hub a help in identifying the spider that made the web.
- A thin layer of threads surrounds the hub. This ***Strengthening Zone*** may be laid down with non-sticky material and looks like an outer part of the hub, often hard to tell from the hub. But with the open hubs, it can be seen better.
- A ***Free Zone*** surrounds the hub. This space is crossed only by the radial threads and no sticky spirals attached. This open space allows the spider to cross from one face (side) of the web to the other. Since the orb webs are usually vertical, this becomes important for the spider to move to both sides and subdue prey.
- Some spirals will also form an arrangement in the center called the ***Stabilimentum***. These silken threads from Aciniform gland silk are added to the hub and neighboring radii in a zigzag manor; sometimes even circular. This structure may help in camouflage, since it is only built by the spiders that always sit in the hub. Other reasons for the stabilimentum is for making the web more conspicuous and therefore less likely to be hit by flying birds. And since the silk of the stabilimentum reflects UV light and pollination insects can be lured to the webs; getting more food. (In our region, *Argiope*, *Cyclosa* and some *Micrathena* usually have stabilimentum in their webs.)

Another structure plays a role in the construction of an orb web, but it is no longer present in the final web; the auxiliary or temporary spiral. Once the radii are formed from the outer edge to the center, the spider makes a spiral thread going from the strengthening zone and the hub and extending it towards the periphery. This auxiliary spiral is made of silk from the Ampullate Minor gland and is attached to the radii to stabilize the half-finished web. The temporary thread serves as a handy guide for the spider when proceeding to laying down the catching (sticky) spiral. After the auxiliary spiral is completed, going out from the hub, the spider goes on to make the catching spiral that is formed from the periphery towards the center. As the spider places these sticky threads in place, it takes up the auxiliary (non-sticky) spirals since it is no longer needed. The final orb web that we are likely to see does not have this auxiliary spiral visible.

There are variations to these parts of the orb webs, but not much variation with the procedure of construction of the webs. It takes place in these three parts:

1. Frame and Radial Threads
2. Auxiliary (temporary) Spiral
3. Catching Spiral

Frame and Radial Threads

The spider starts with a horizontal thread by simply exuding a silk thread into the air while sitting on an elevated site. This action is sometimes said to be that the spider shoots or throws out the silk. Actually, a hind leg is used to pull the thread from the anterior and median spinnerets. This thread is made from silk of the Ampullate Major gland. This thread is carried away by wind, breeze or a slight drift until it gets caught on some object and the spider feels its attachment. Most orb webs are made at about dusk when the winds of the day have mostly calmed. Next the spider reinforces this initial thread by walking back and forth on it, laying down more silk. A thread is then fastened in the center of a loose horizontal thread. The spider then lowers itself down to an attachment on a substrate below; this could be a branch, a board or even the ground. Pulling tight, the spider creates the hub. Going back up to the hub and then to the sides, the spider forms the surrounding frame or scaffolding. Repeated trips to the hub and on to the outer edges makes the various radii. The radii are at spaces that allow the spider to step from one to another; they are not sticky. Small spiders tend to make webs with more radii than larger spiders. The spider decides at the hub in which sector the next radius is to be placed. As soon as it has completed a radial thread and has arrived back at the hub, it briefly tugs on all the finished radii with its front legs; appearing to measure the angle between the radii. Such angles are astonishingly constant. Radii are constructed in a progression in different parts of the web; not usually in one next to another.

There is a consistency in the number of radii in the web for the different spider species. I regularly use this as a characteristic for identifying their webs.

Auxiliary Spiral

While the spider is still busy building the radii, it is also interconnecting these threads by a few narrow circles in the center of the web. This hub region is then extended into the region known as the strengthening zone by further peripheral turns. After construction of the radial threads has been completed, the strengthening zone is continued into a wide spiral towards the periphery; forming the

Web construction

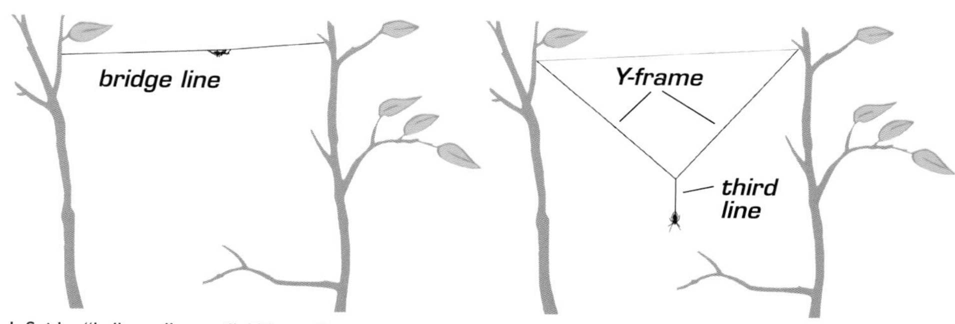

1. Spider "balloons" out silk. When silk gets snagged, spider tests the hold and then reinforces it with more lines.

2. Spider trails a loose second line, fixed at either end of the bridge line. A third line is made in the middle.

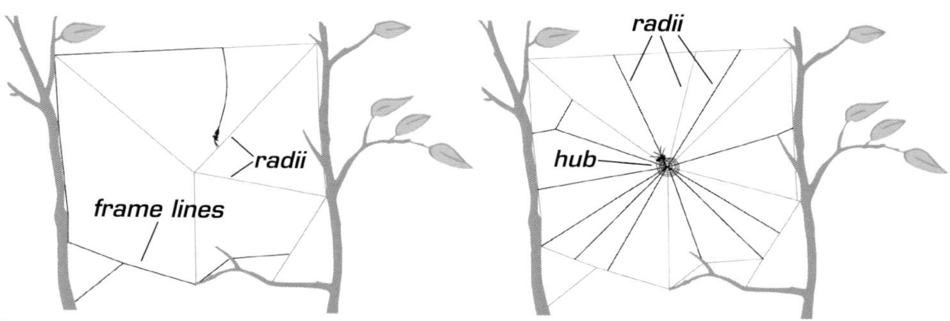

3. Spider returns to frame and spins outer anchoring threads and more radii (spokes).

4. Once the frame and radii are completed, a tight, spiral hub is spun.

5. Next, an auxiliary or temporary spiral stabilizes the web and creates "stepping stones" for spinning the sticky snare.

6. The spider now retraces its steps, eating the temporary threads and replacing them with sticky threads.

auxiliary spiral (may also be called the temporary spiral). Although the auxiliary spiral has only a few turns, it ties the radii together and helps stabilize the half-finished orb web. The auxiliary spiral thread also serves as a handy guide for the spider when it comes to laying down the catching spiral. Without this auxiliary spiral (non-sticky), the spider can cross the radii only awkwardly, since each radial thread would sag greatly under the spider's weight. After the auxiliary spiral is finished, the spider pauses for a moment before it tackles the last step of the orb-web construction; the catching spiral.

Catching Spiral

Laying down the catching spiral (also known as the sticky spiral) is a remarkable accomplishment of coordination. The catching spiral thread is made from silk of the flagelliform gland attached to the posterior spinnerets. While this thread is exuded and put in place, it has sticky glue globules added. This sticky material comes from the aggregate gland; also attached to the posterior spinnerets. Meanwhile, silk of the piriform gland from the anterior spinnerets forms the points of attachments between the spiral and the radii.

The auxiliary spiral begins at the hub and ends well before reaching the peripheral framework. When laying down the sticky spiral, the spider starts at the periphery and uses the auxiliary spiral as a guide line. While crossing the radial threads in a spiral fashion, it constantly fastens the sticky thread to each radius. In this process, both front legs and the hind legs play an important role. One front leg reaches for the next radius and touches it with a quick brushing movement, apparently to ascertain its position. At the same time one of the fourth legs pulls the thread out of the spinnerets and dabs it against the radius. While doing so, the spider keeps the distance between successive turns about equal.

As the construction of the sticky spiral proceeds, the auxiliary spiral is simultaneously taken down. In the finished web, only the remnants of the crossing points can be recognized. The catching spiral is not continued into the center of the web, but terminates shortly before it. This results in a rather open space between the catching area and the hub (termed the free zone) that allows the spider to move from one side of the web to the other. There are a few exceptions to this; some catching spirals will extend into the hub. The sticky catching spiral is by no means a continuous spiral thread. It often reverses its course from clockwise to counterclockwise, and vice versa. Since the position of the hub is slightly off center, that is, shifted towards the upper part of the web, most of the sticky spiral is confined to the lower half of the web.

Laying down the sticky catching spiral is the most time-consuming stage of the orb web's construction. Whereas building the framework and the radii take only about five to ten minutes, the sticky spiral requires twenty to thirty minutes. That the entire orb web in all of its complexity is created in one-half to one hour is astounding. The total length of thread created measures several meters long.

It is also remarkable that the construction of the orb web is controlled only by the sense of touch, without any visual feedback. The orb-web spiders tend to have poor eyesight and often the web is made in darkness shortly after dusk.

The ultimate purpose of any spider web is the capture of prey and the orb web is certainly no exception. Its highly geometrical construction suggests a special effectiveness and economy as a trap. As we have just seen, the orb web is built with a minimum of material and time. The tightly strung radial threads converging in the web's center give the spider two advantages: they transmit vibration signals from the periphery to the center, where the spider usually sits, and they provide direct and quick access routes. The web can be regarded as an extension of the spider's senses. It has been shown that the longitudinal vibrations of the radii are important, since they follow a specific direction and are hardly changed. The spider's first reaction to a fly that has hit the web is to jerk the radial threads several times with its front legs. Only when the prey has been located in relation to the center does the spider rush out on a directed course; in most cases exactly on that radial thread leading straight to the prey.

The spider is able to avoid being caught in her own web in three ways; she moves about the web by staying on the non-sticky radii. Also, spider feet have an oily coating that keeps them from being caught. And on their feet, orbweavers have a third claw (most spiders have two), that they use to hold on. Despite the fact that they have poor eyesight, they do not get trapped by their own snare!

The radii serve as communication channels not only during prey capture, but also during courtship, when the male approaches the female's web. Usually the smaller male sits at the edge of the web and rhythmically plucks the threads, apparently in order to not be mistaken for prey. The mechanical properties of the sticky spiral, especially its great elasticity, are also very important; it can be stretched several times its original length without breaking. This is of course advantageous when a vigorously struggling insect is trapped in the web. Aside from these mechanical stimuli, the web can also carry chemical signals, which stem from the spider's secretion of pheromones. It is known that female's empty web may elicit courtship behavior of the male. The carrying of chemical stimuli is not restricted to orb webs alone, but applies also to the draglines of wandering spiders as well.

The hub is usually slightly off-center; it lies above the geometrical center of the web. Apparently this serves a purpose: the spider sitting in the center can run a little faster downward than upward in the web. This means that prey in the larger lower half can be reached as quickly as in the smaller upper half of the web. And the insects are more likely to hit the larger section of the web.

Most orb webs have a vertical orientation in space, or to be more precise, they are slightly tilted. The spider sits almost invariably on the "underside" of this tilt, often in the hub. Nearly all of the spiders in the hub sit inverted; heads facing down. In case of a disturbance or danger, the spider can simply drop to the ground on a safety thread; a response that is typical of many orbweavers.

In nature, orb webs must be replaced often, because they quickly become damaged or destroyed by wind, rain or early morning dew. As a rule, many build a new web each day. Only the previous framework may be used again for the new construction. Taking down the old web is a rather systematic procedure (see illustrations below). While walking out from the hub, the spider destroys three to five radii together with the crossing sticky threads, thus producing an open section in the orb. After one-half to one hour, the web is reduced to the frame and perhaps a few radii. At this point, the spider starts building a new web (left illustration below). Some species of orb weavers practice a much faster technique of taking the web down in about one minute. Cutting the lateral frame

Web Destruction

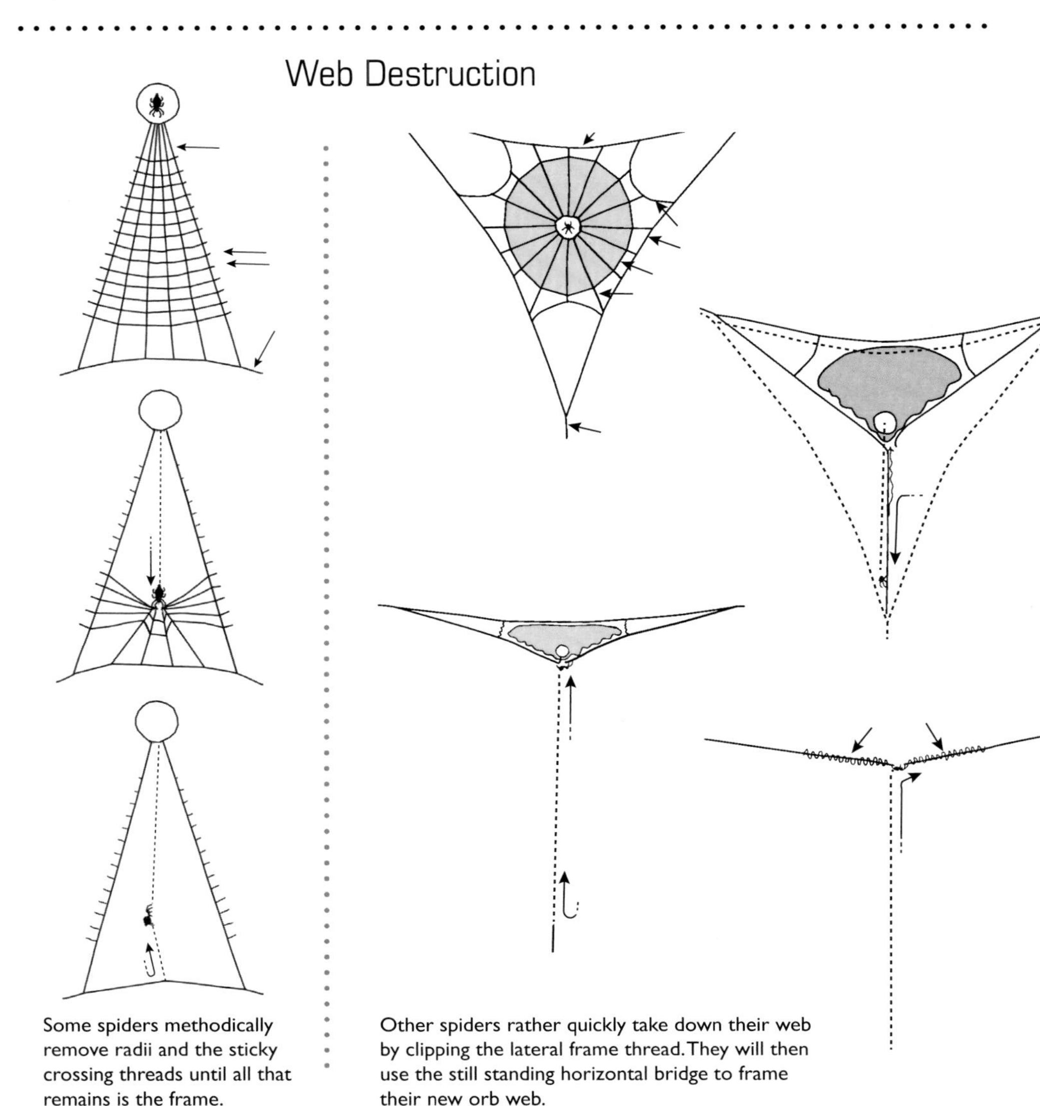

Some spiders methodically remove radii and the sticky crossing threads until all that remains is the frame.

Other spiders rather quickly take down their web by clipping the lateral frame thread. They will then use the still standing horizontal bridge to frame their new orb web.

threads makes the entire web collapse so that only the horizontal bridge thread will remain. The new web construction begins here by first establishing the basic Y-structure and the new hub (right illustration). The material of the old web is eaten by some orb weavers, other species simply throw it away; and some use it to wrap their egg cocoons.

If the web is only slightly damaged, as may happen after catching small prey, the spider may or may not mend the hole. In general, repair does not consist of reconstructing the original design, but of patching the holes to make the web stable again.

Web Watching Tips

For those who wish to get more acquainted with the webs of late summer, I invite you to put on boots and walk among the dew-covered plants of roadsides, fields, meadows and swamps. In this region, we have a period of time of about a month from mid August to mid September when conditions are best. A clear morning often has mist and dew in the early hours and is the time to see plenty of webs. But I have found foggy mornings are outstanding as well. Bring along a camera and close-focus binoculars. Walk towards the rising sun. Webs are best seen when back-lit by the new dawn. Usually the winds are calm at this hour and the morning light is diffused; great conditions for photography as well as observing.

Orb webs dominate the scene, but sheet webs, funnel webs and cob webs also abound. I have regularly found an abundance of funnel webs in mowed fields and roadsides while Dictynid cobwebs are on many standing stalks and a plethora of sheet webs are in the Leatherleaf plants of swamps and bogs. The dew cover on the silk threads makes the webs easier to see, but if there were no dew, the webs would still be common here. I prefer to use these droplets formed by dew or fog

Dewy or foggy mornings in late summer are the best times for observing spider webs.

Fog can be a web watcher's best friend. The webs stand out much better in these conditions.

to enhance the web sites.

Though these dew-covered webs may be a treat for us to see, I'm not so sure that the spiders appreciate them. As I walk in this scene during the early hours, I find many webs with no spiders visible. It appears as though the dew has sent them to the sidelines. Others remain in the web, usually the hub, and wear the same wet attire as does their web.

These wet web walks are best taken slowly, but don't go too slow. The rising sun and rising temperatures and winds on these late summer days will disrupt the dew quickly and I find that by about 8 a.m., the morning web walking and watching has lost its dew and moved on to the next phase.

Webs can be found at other places as well and once we have acquired the eye for how to see the webs, we may search for webs, orbs or other types, in our yards and in the woods. Though the dew cover here will vary from out in the open, we know what to look for. It is still best to search for webs in the morning.

As the day warms, winds pick up and often, the webs are destroyed. As we learned earlier in the book, several species of spiders even take down their webs and eat the silken threads later in the day. New webs may be constructed again the following evening.

While many spiders make the large webs that we see out in the open fields and meadows, others are in the woods. I have found it best to walk towards the rising sun to see the webs, even if it is a walk in the woods.

Several species make webs along lakes, ponds, swamps and river shores. Frequently they use cattails and other aquatic plants as the substrate, but also check docks, boats or shoreline trees; especially alders. When walking among the webs we'll learn not only who makes the web, but where to look for the resident spider; whether in the web or a nearby retreat.

Webs certainly are beautiful, but they are also marvels of construction. The protein silk material from the internal glands of spiders is formed into amazing structures that are strong and elastic; and sometimes very long lasting.

Though I do most of my web watching and photography in the dew-covered early mornings, this is not the only time to find them. Another excellent time to go out for a web walk is at dusk or shortly after. Many orbweavers form new webs each evening and so by getting out at this time, we can see the webs being made and possibly the spider in it. The owners of cobwebs, sheet webs and funnel webs are often more likely to be seen in their webs during the evening as well. I like to search the walls of buildings. I've had good luck seeking webs on the siding of houses and garages around dusk. It is still warm enough on the late summer evenings that the spiders are active. There is prey aplenty, especially moths, but also flies, mosquitoes, gnats, midges and more.

Whether we appreciate them or not, spiders and their webs are around us constantly. And whether it is late summer for photography or any other time, we are likely to see these critters that dwell amongst us. Let's learn more about the webs and the spiders who make them.

Cob Webs

Cob webs, sometimes called irregular webs, can appear to be a haphazard group of threads. The spider sits inverted in the center. These webs are found both indoors and outdoors.

Pholcidae are nearly always indoors. Dictynidae are outdoors. Theridiidae can be found both. All of these can be found from Spring until Fall; the indoors ones may be seen throughout the Winter.

Families of Cobweb building spiders:
Theridiidae: Cobweb Weavers (a.k.a. Comb-footed Spiders)
Pholcidae: Cellar Spiders
Dictynidae: Meshweb Weavers

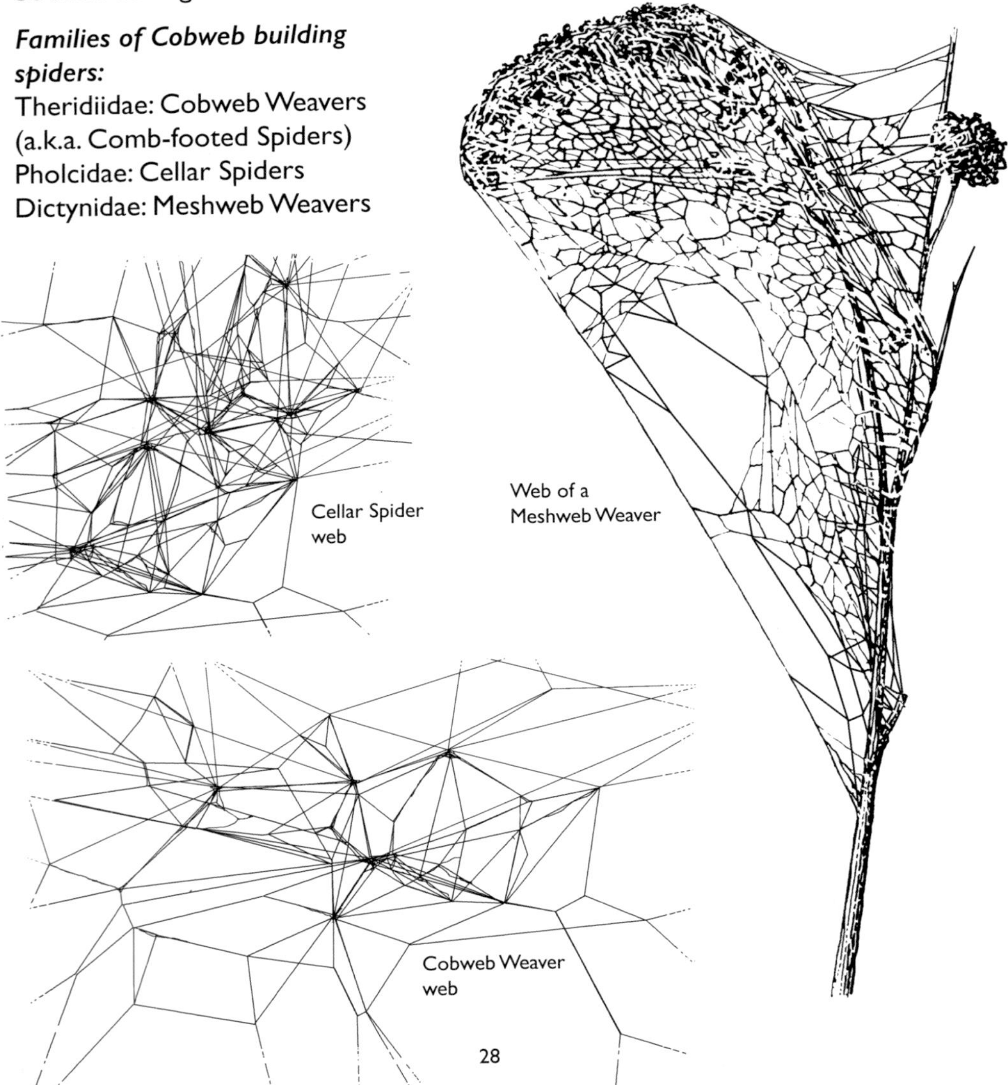

Cob Webs

Cob Webs

Southern Black Widow & Northern Widow — *Family Theridiidae*

Web—Irregular Cobweb

Like other members of the family Theridiidae, the southern black widow will make an irregular web; often this snare is referred to as a cobweb. It is made of strong and coarse silk and most commonly near the ground around tree stumps, wood piles, under stones or holes in the ground. But they can also be found in garages, barns, storage buildings, outhouses and other human structures. Like many of the cobweb weavers, the spider sits inverted in the center of the web. Being shy, when disturbed, they tend to retreat into a nearby shelter.

Southern Black Widow

Observations: I have probably seen more of the Widows; both Southern and Northern, while they were sitting in the irregular web, than anywhere else. Though there are many reports of webs found in buildings, I think that nearly all of my sights have been on or near the ground; usually among rocks or in holes.

Southern Black Widow & Northern Widow *Latrodectus species*

Spider

The Southern Black Widow (*Latrodectus mactans*) has a shiny black abdomen with a red hourglass-shape marking on the underside. Since it hangs in the web upside down and since the webs are often low near the ground, this distinct marking is easy to see. The topside of the abdomen is black. The closely related Northern Widow (*Latrodectus variolus*) is more likely to have red spots on its back and the hourglass pattern on the underside is separated into two parts. It is found in the eastern U.S. from Florida to southern Canada. Females of both species are the largest of all cobweb weavers at about a half inch long (12 mm).

Cob Webs

Common House Spider | *Family Theridiidae*

Web—Irregular Cobweb

Web appears to be a mass of confusing threads. They are usually less than a foot in diameter. Webs are commonly seen indoors and these are the cobwebs that we are likely to find indoors–in corners, window sills and heating ducts. Though appearing to be a mess to us, the spider constructs it to catch prey. Some of the lower threads are sticky, but most are not. Spider sits inverted in center of web. Female and male and egg sacs may be in this web at the same time, but it is usually the female that we see. The webs may last for a long time indoors and the spider will survive all year. Outdoors, they thrive often near buildings in summer into early fall, but they do not survive winter outdoors.

Observations: I find these webs abundant on the outside of buildings, among the siding and corners or under eves. During the daytime, the webs are very obvious, but not the spider. They hide in the cracks of these sites. But returning after dusk, you may find the spiders in the webs. Exoskeletons of previous meals may be in the web with them.

Common House Spider *Parasteatoda tepidariorum*

Spider

About 6 to 8 mm long with a legspan of about 15 to 20 mm. Legs are long and thin; body is pear-shaped. Colors range from dirty white to brown to near black.

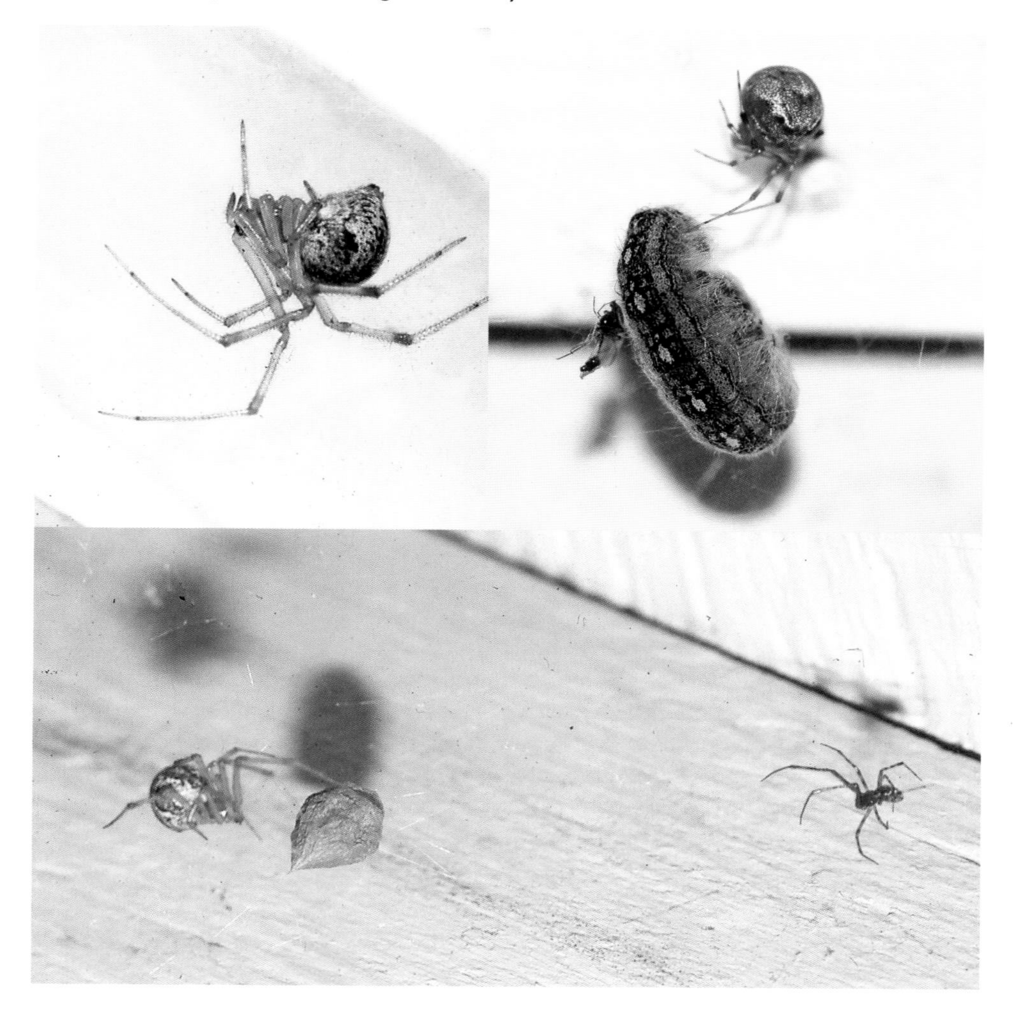

Cob Webs

Northern Cobweb Weaver *Family Theridiidae*

Web—Irregular Cobweb

Though the web is an irregular cobweb, it appears to be more organized than most. Web is small. The upper part of the web is an irregular sheet with threads attached to a substrate above. Below is a series of sticky threads attached to a lower structure. Webs may be indoors or outdoors. The indoor webs allow the spider to winter indoors. Spider usually in center of web; mostly at night.

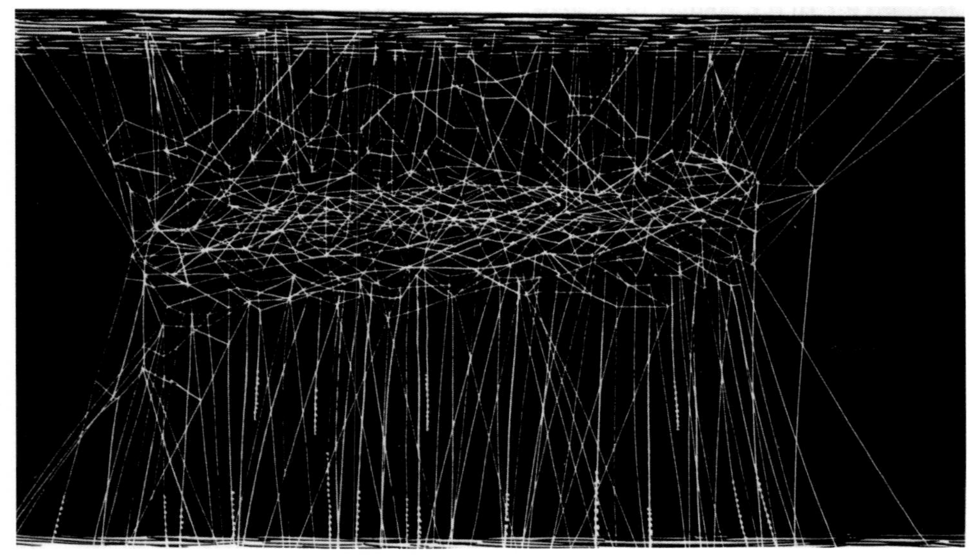

Observations: I have found this web both outdoors and indoors. Outdoors the webs may be in low vegetation, under bark or rocks. I find many at the lower part of tree trunks and on the outside of buildings. Indoors, they can be found in the corners of rooms, basements and garages. Practicing amazing patience, I have seen them staying in their web without moving for many days. Outdoors, they hide in the day in cracks or slight openings, but go into the web at night.

Northern Cobweb Weaver *Steatoda borealis*

Spider

Small; body length of 6 to 8 mm, with a legspan of 12 to 15 mm. Abdomen oval in shape; glossy dark with light line around the edge and in the center. Spiders are usually seen in the web in an inverted pose.

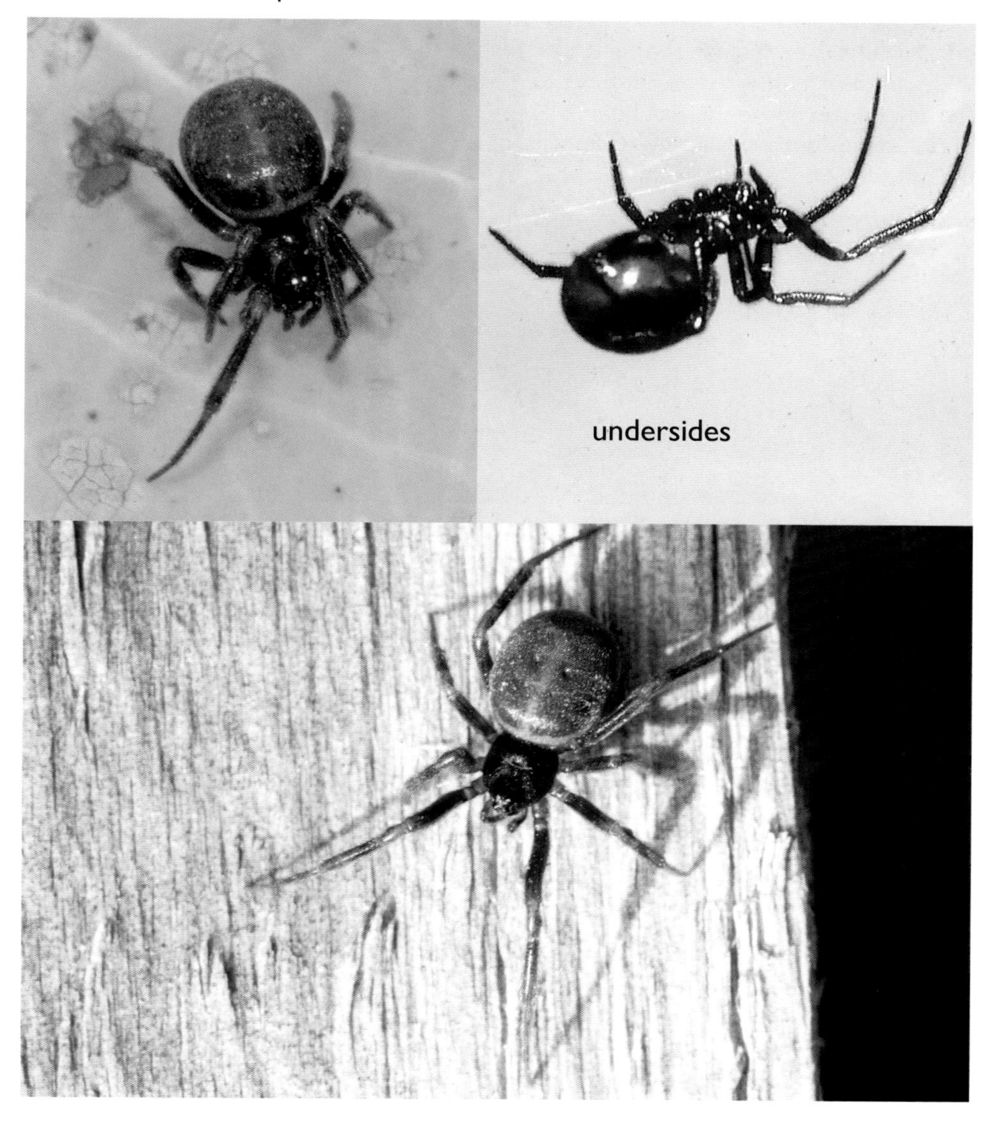

Cob Webs

Leafy Cobweb Weaver | *Family Theridiidae*

Web—Irregular Cobweb

Web appears to be a haphazard mess of threads. But the large number of threads of this web (usually less than a foot in diameter) are in some organization. Spider sits inverted in the center. Webs are nearly always seen outdoors and several feet above the ground. Webs are in shrubs, small trees and on the outside of buildings.

Observations: I have found many of these webs in shrubs and small trees. Not as common as some other cobwebs on buildings, they can also be found here. Female and male may be in the same web. Not only do they have their egg sacs in the web, but for a while they will tolerate the young living here. I have seen the young in the web in the daytime. Some say that this is as close to a social spider as exists in our region.

Leafy Cobweb Weaver | *Theridion* species

Spider

Small; 3 to 4 mm long with a legspan of about 6 to 8 mm. All *Theridion* species have an oval-shaped body with variable markings (see photos below). *Theridion pictum* has a bright red arrowhead pattern atop the abdomen while *Theridion frondeum* has a pale wavy band down the center. Spiders sit inverted in the web and may be seen in the daytime.

Cob Webs

Longbodied Cellar Spider *Family Pholcidae*

Web—Irregular Cobweb

Web is often small and may be quite hard to see. They have fewer threads in the web compared to other cobweb weavers. Nearly always indoors. They have adapted to life indoors so much that the spiders remain active all winter, even continuing to breed throughout the year. Females hold the egg sac in their jaws. Webs are found along walls and in corners, near ceilings or, as their name implies, frequently in cellars, basements and garages. Outdoors, they may be in caves or along rock cliffs. The spider with its long legs hangs inverted in the web.

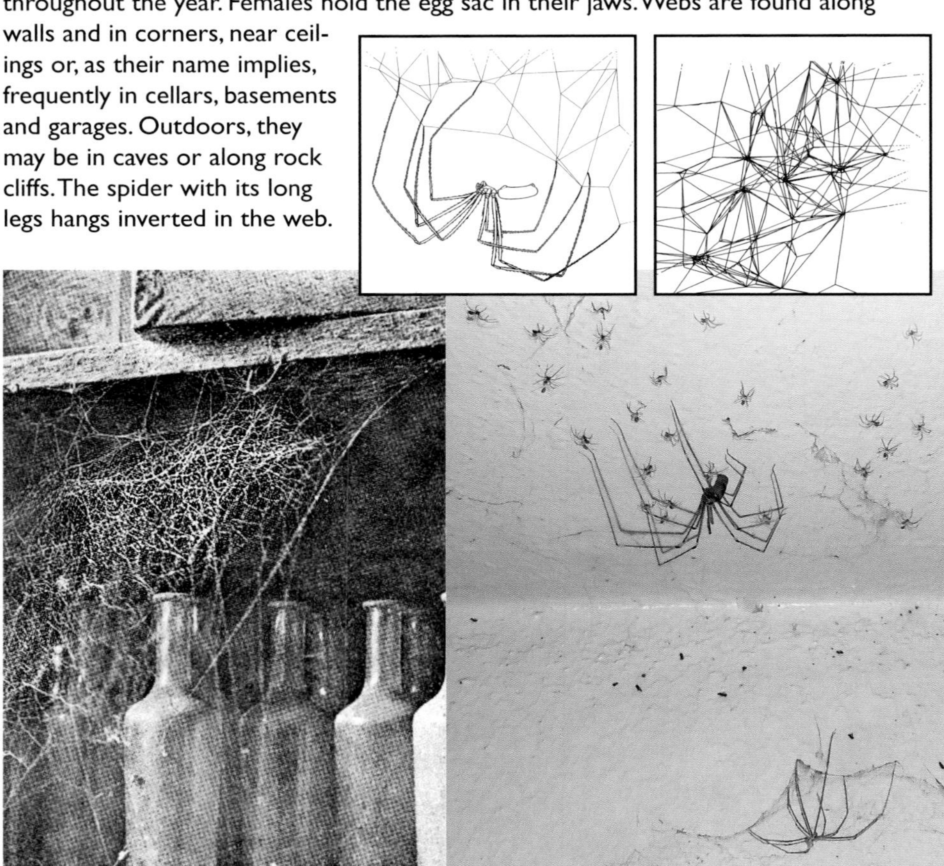

Observations: I have seen these spiders in their webs indoors often, but only once did I find them in a small cobweb outdoors; and that was near a building. Spiders can be found all year and when I need live spiders for a program in the winter, this is the one that I look for. They can also be seen in the web day or night. When disturbed in the web, they shake their body in a rapid movement so that appear to be just a blur.

Longbodied Cellar Spider *Pholcus phalangioides*

Spider

Body is elongated about 7 to 8 mm long with extremely long legs making a legspan of 35 to 50 mm. Such long legs accounted for the former name of this spider—the daddy longleg spider. This is confusing since the true daddy longlegs is a completely different arachnid and not even a spider. Bodies are usually light in color; gray. Like other cobweb spiders, they are often immobile.

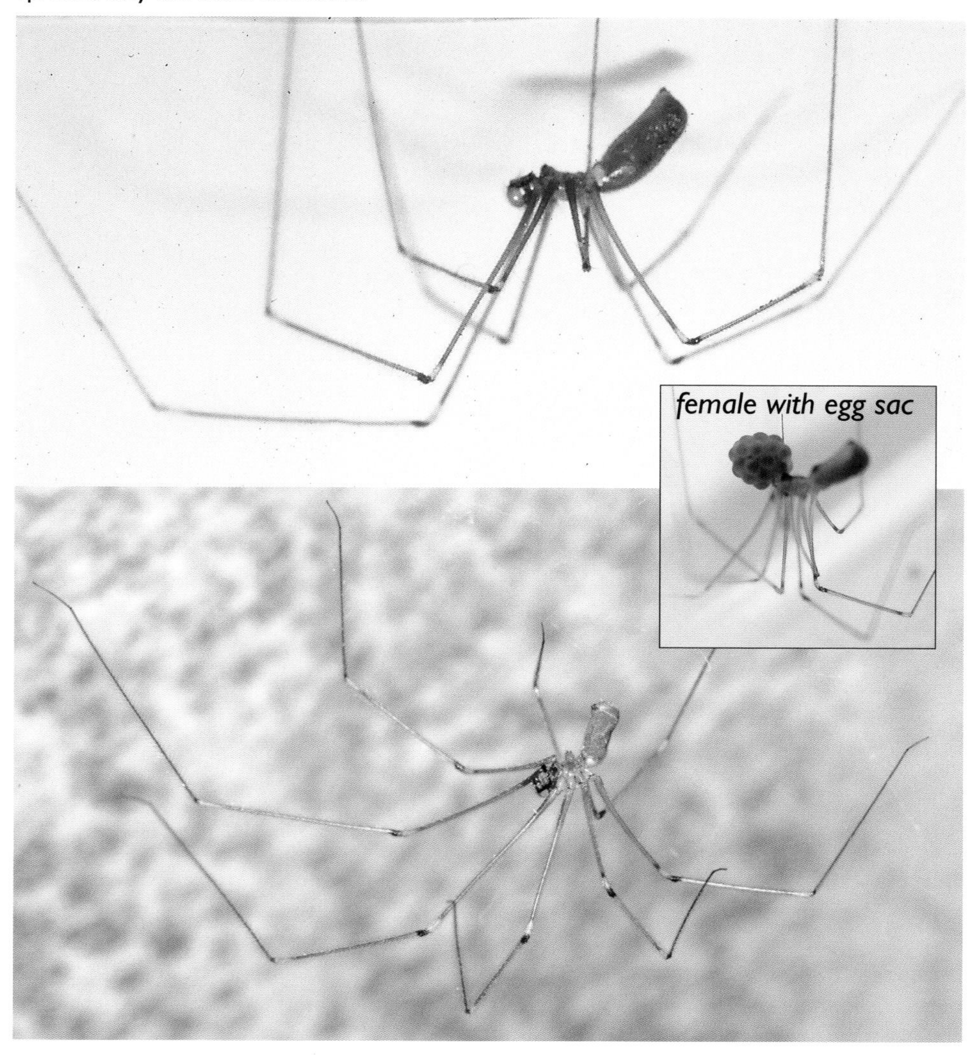

female with egg sac

Cob Webs

Meshweb Weaver — *Family Dictynidae*

Web—Irregular Cobweb

Webs of this group of spiders are perhaps the smallest of the cobweb weavers in the region, but the webs may be the easiest to see. Webs are constructed on the tops of plants in fields, meadows and along roadsides. Living plants are used, but also frequently on the tops of dead standing plants late in the season. Webs are irregular and look like a mess of threads. Spiders are tiny, but are usually in the webs. A close look will reveal them.

Observations: A dewy summer or fall morning, will make many webs easily visible. One of the most common are the Dictynid webs. I have never seen their webs indoors. The site of choice is the tops of plants, maybe up to three feet high. I have never seen these near the ground though they have been reported to be among rocks and other debris. Spiders are very small, but looking carefully among the threads can reveal them. They remain active quite late into the fall.

Meshweb Weaver *Emblyna annulipes*

Spider

Small; body only about 4 mm with short legs; legspan is from 5 to 6 mm. The oval abdomen is mostly dark. Spider sits inverted.

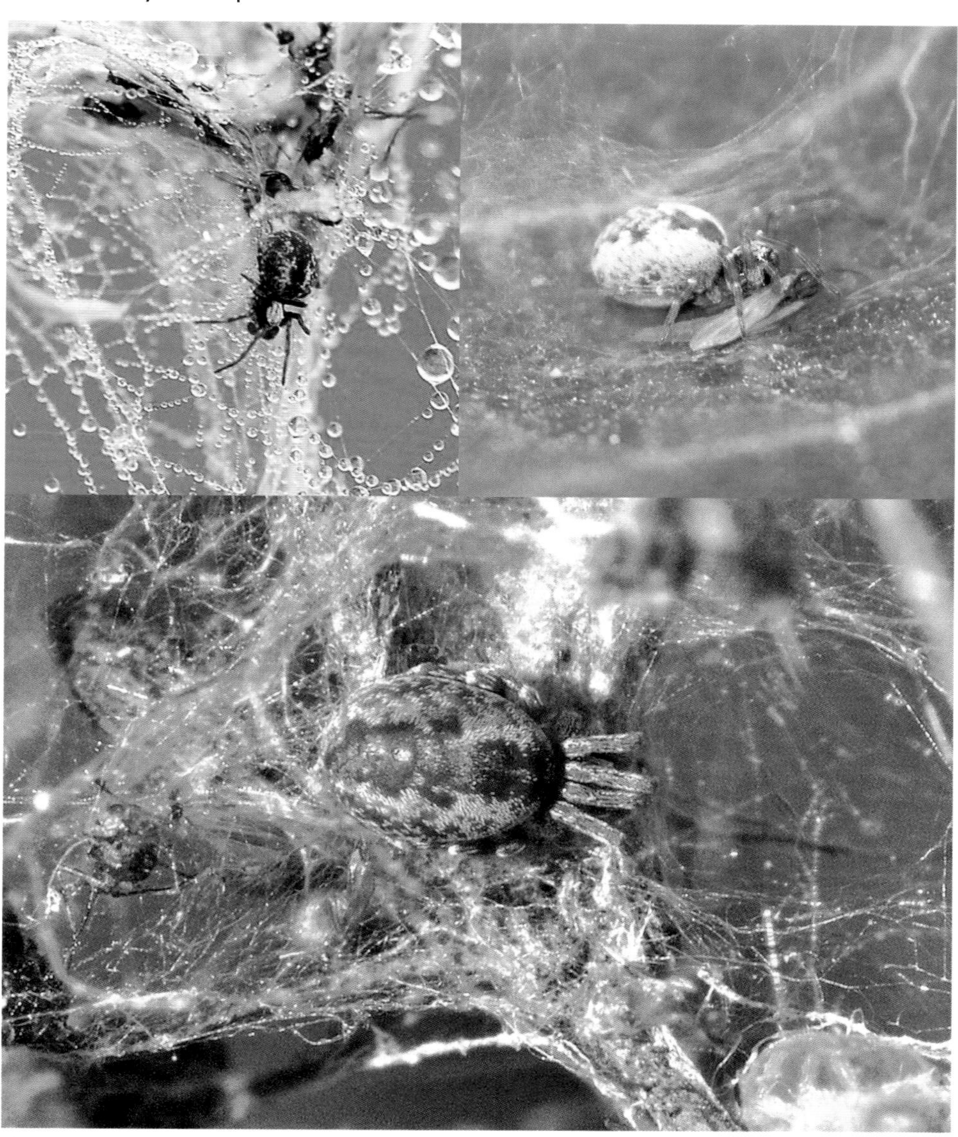

Sheet Webs

Sheet Webs appear to be either bowl-shaped, dome-shaped or flat. The spider is beneath the flattened part of the web.

Outdoors these webs are on shrubs, trees and often in swamps. Webs can be found from spring to fall, but best in late summer.

Families of Sheet web building spiders:
Linyphiidae; Sheetweb Weavers
Hahniidae: Dwarf Sheet Spiders

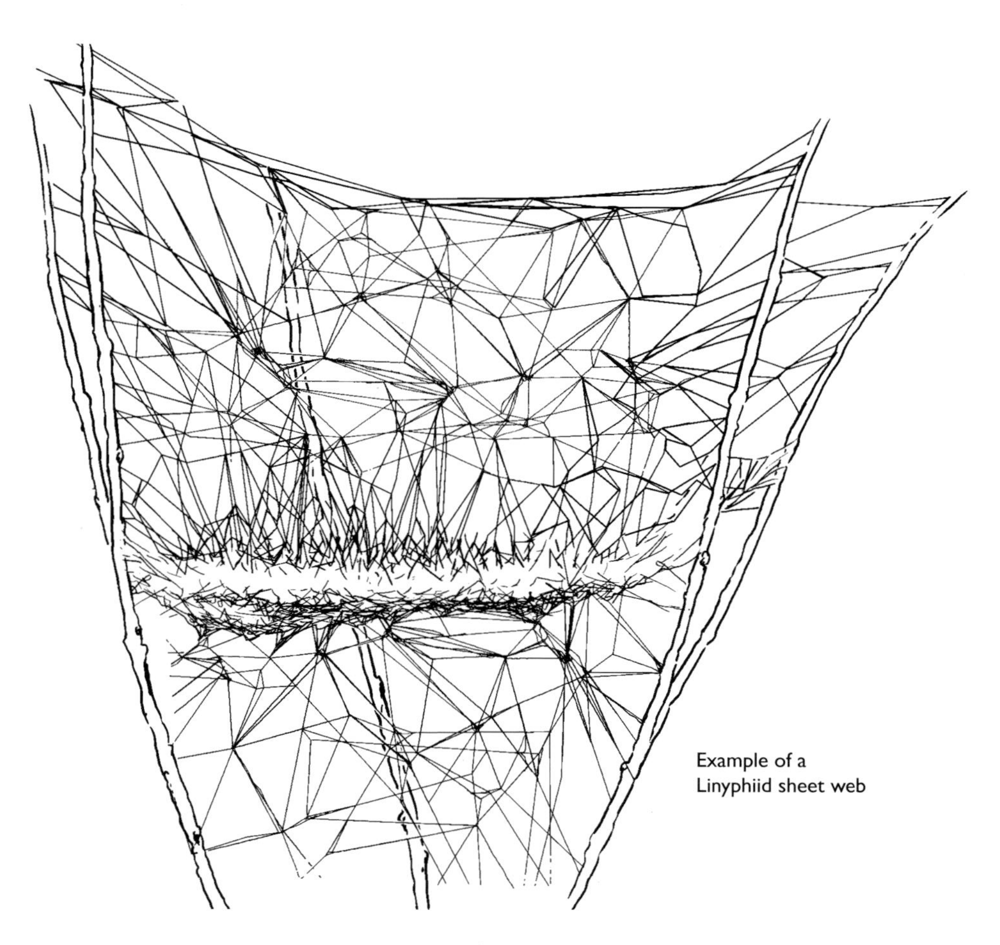

Example of a Linyphiid sheet web

Sheet Webs

Sheet Webs

Bowl and Doily Weaver | *Family Linyphiidae*

Web—Sheet Web

Webs are usually a few inches wide and tall; but can be up to a foot. The main part of the web is cup-shaped (bowl) with a flat part (doily) below, with many threads above and below. Spider sits inverted under the bowl. Insects hit the top tangle of threads and fall into the "bowl." The spider then captures the prey. Webs can be found from spring, throughout summer and into fall.

Observations: I find these abundant webs on dew covered mornings. They last for many days or weeks, but can be seen best when covered in dew and backlit. I have found them in fields, along roadsides and at edges of woods, but they appear to be most common in shrubs; especially evergreens and in swamps. Leatherleaf plants in the bogs often host dozens of these webs. Spiders remain in the web; no retreat. Members of this group are common late-season spiders; often ballooning and traveling in fall.

Bowl and Doily Weaver *Frontinella communis*

Spider

Body is small; 3 to 4 mm with long legs, making a legspan that can be 9 to 12 mm. The oblong abdomen is dark with white markings on the sides.

Sheet Webs

Filmy Dome Spider | *Family Linyphiidae*

Web—Sheet Web

Web is dome-shaped; an inverted bowl, if you will. Small, they are only four or five inches in diameter. The spider sits in an inverted pose under the top part of the dome. Threads above the dome help tangle victims and they fall onto the top of the dome. The spider then reaches up to subdue the prey. Webs are mostly in the lower branches of trees in a deciduous forest.

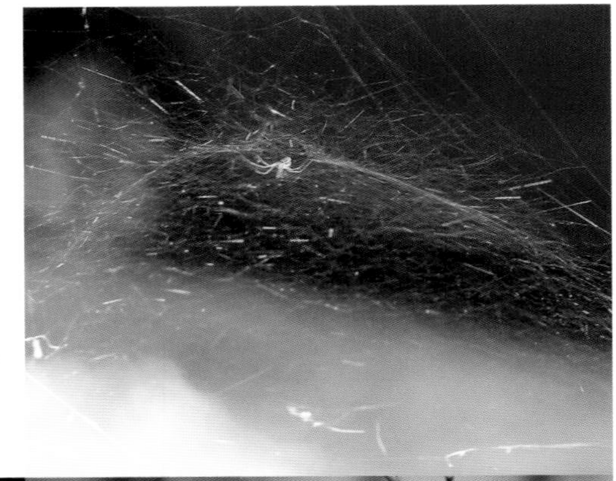

Observations: I find almost all these webs in late summer and fall in the lower branches of the woods. They are not as abundant as the bowl and doily webs. Like most webs, the domes are easiest to see in the morning with a dewy coat. Dew is not as thick in the forest, but I have found that walking slowly towards the rising sun and looking closely at the branches, they can be found. Spiders do not leave the web; no retreat. Web may last for days.

Filmy Dome Spider *Neriene radiata*

Spider

Small body: 4 to 6 mm. Long legs with a legspan of 15 to 20 mm. The oblong abdomen is dark with patterns of light and dark on the sides.

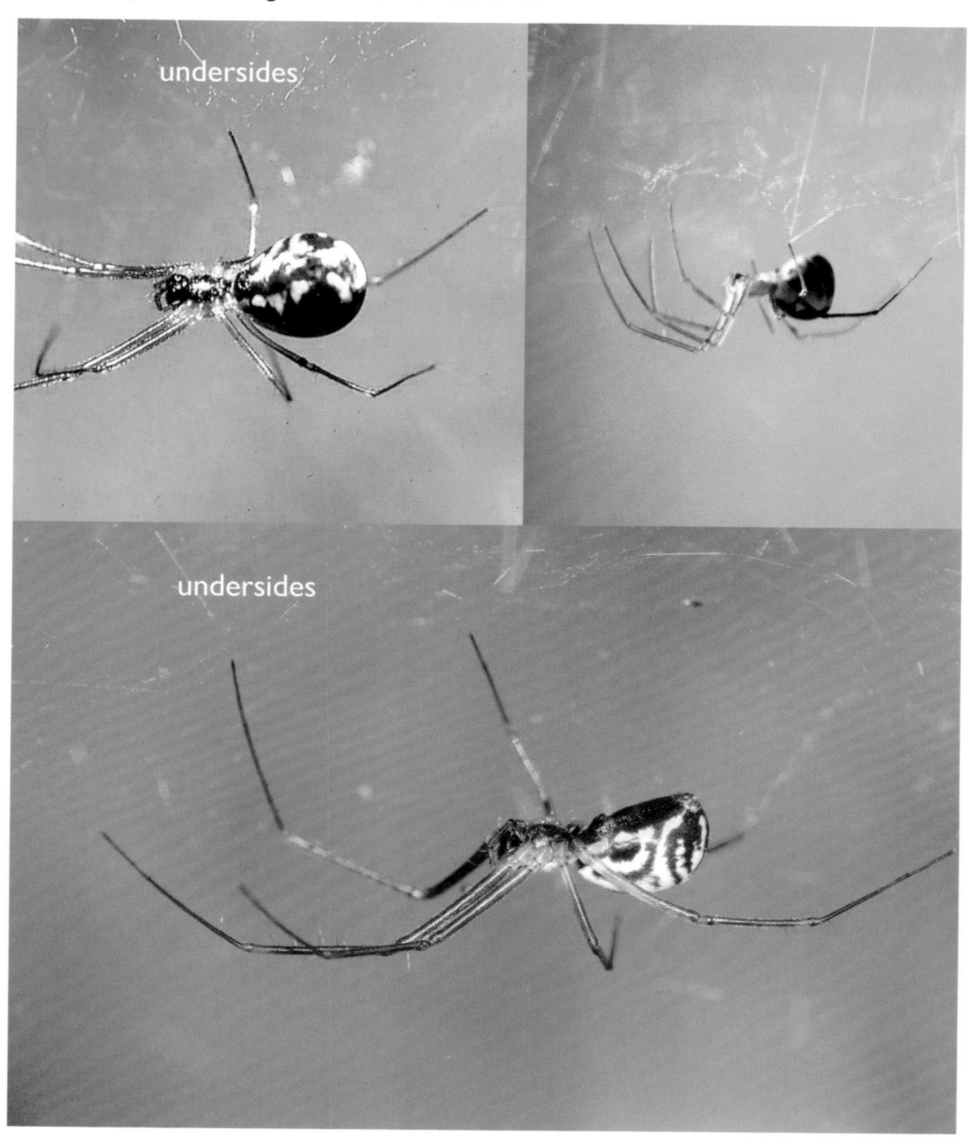

Sheet Webs

Hammock Spider | *Family Linyphiidae*

Web—Sheet Web

Webs are longer than most sheet webs, up to eight or ten inches. These webs are also more flat, more like an actual sheet, and often like a hammock. Usually in the lower branches of trees, they can also be found on buildings, or under eaves. Spiders are in the web, but frequently in an adjacent curled leaf retreat. Webs remain active quite late in the fall.

Observations: I'm sure that this spider is common all summer and fall, but I usually do not see the webs until about the time of the leaf drop in October and through the next few weeks. Despite the chilly weather, I find them easily in fall. Since many Ironwood trees keep curled brown leaves, the spider will use these as retreats and build the flat web out from here. On first glance, the spider is often not seen, but I find her in the leaf retreat.

Hammock Spider *Pityohyphantes costatus*

Spider

Body is 5 to 7 mm long with a legspan of 14 to 18 mm. The oval abdomen is light to gray on the sides with a dark pattern in the center.

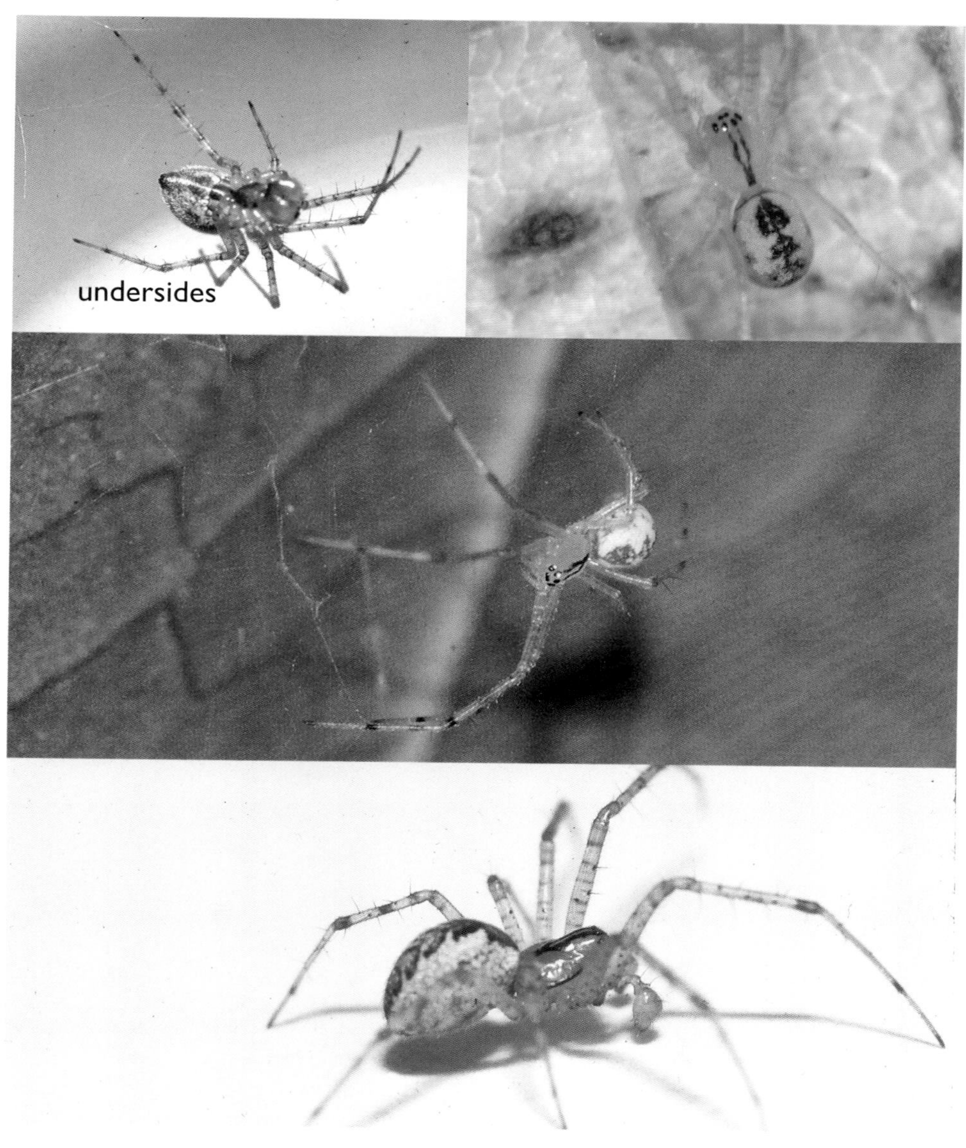

Sheet Webs

Dwarf Sheet Spider | *Family Hahniidae*

Web—Sheet Web

These tiny spiders make a very small flat web, usually no more than two inches across. Webs are on the ground in forests, meadows, fields, gardens, lawns and often near damp places; in slight depressions as formed by footprints. Spider sits under the web or towards the edge. They make no retreat.

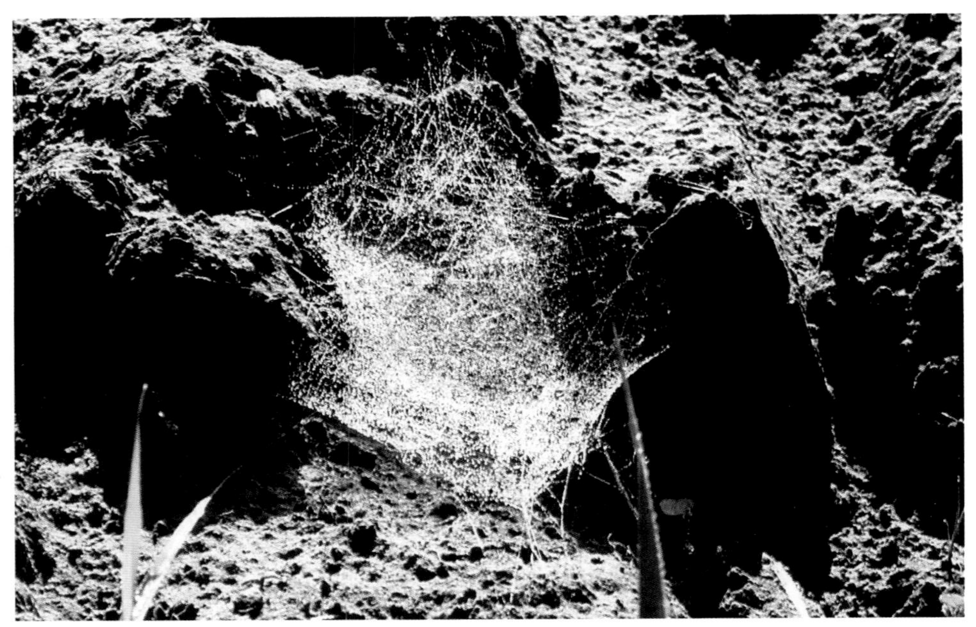

Observations: These miniature spiders are hard to see, but with a dew cover their small flat webs can be seen on the ground. I have found many right in the yard, in the small depression formed by a footprint. The spider is under the web.

Dwarf Sheet Spider | Hahniid spiders

Spider

Body is tiny; 1 to 3 mm with a legspan of maybe twice that. The abdomen is brown with lighter chevrons on the back. This family of spiders has the six spinnerets all arranged in a straight line; giving the sometimes used common name of "comb-tailed spider." In the family Hahniidae, only the genus *Neoantistea* builds webs.

Funnel Webs

With a hole in the center or slightly to the side, these webs resemble a funnel. Webs are usually on the ground. Spider sits at the entrance of the openings.

Webs can be found from spring to fall; most in the mid to late summer. Some spiders come indoors in fall and these webs could be with us all winter.

Families of Funnel web building spiders:
Agelenidae: The Funnel Weavers (most common)
Amaurobidae: The Hackledmesh Weavers

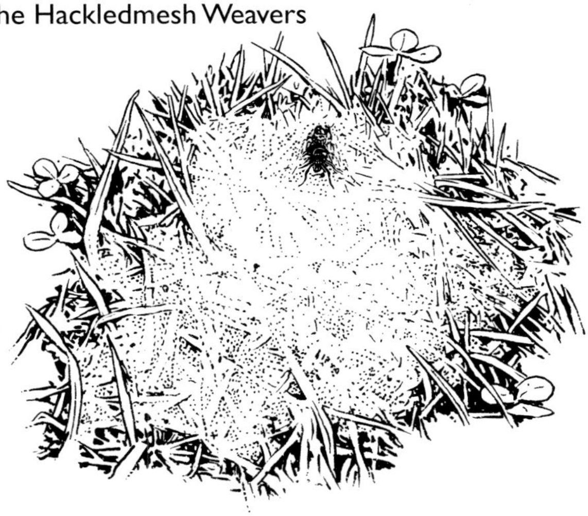

Funnel web of an Agelenid Grass Spider

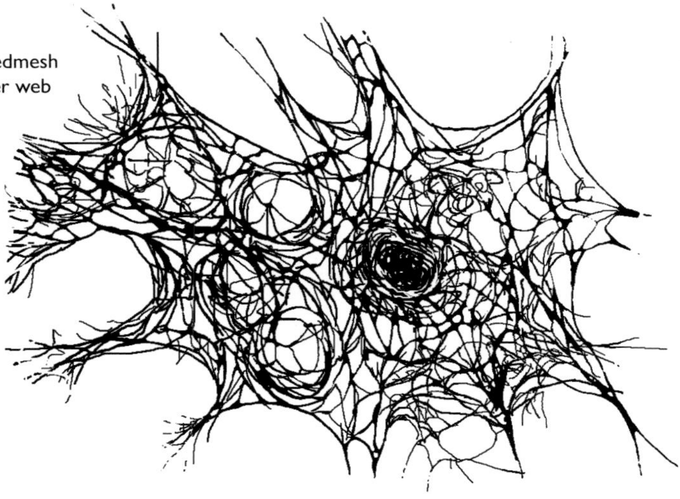

Hackledmesh Weaver web

Funnel Webs

Funnel Webs

Grass Spider | *Family Agelenidae*

Web—Funnel Web

This web gets its name from the funnel-shaped hole in the center or to the side of the web. The rest of the web is often flat. Frequently the webs are on the ground. Spider uses the hole as a retreat or hiding place. Most of their hunting is done at night. Web is not sticky but if prey gets on the web, spider rushes out to subdue it. Many threads around the outside help to force the insects into the web. Webs can be large; up to a foot in diameter. They last for weeks. Webs are abundant outdoors in summer, and with the cold weather, some come indoors.

Observations: I find these webs in huge numbers on the dew covered mornings of late summer, in fields, on lawns and along roadsides. They are especially easy to find in fields or lawns that have been recently mowed. Though late summer is the peak in numbers, I also find webs in spring; the ones that have overwintered. Small webs of younger spiders can be found in summer.

Grass Spider | *Agelenopsis* species

Spider

Bodies are 10 to 17 mm long with a legspan of about 25 to 35 mm. Abdomen is brown or gray, but it is the carapace that has dark and light stripes. Members of this family have long spinnerets that look like a pair of tails. They can be confused with wolf spiders that do not make webs and have shorter spinnerets.

Funnel Webs

Barn Funnel Weaver | *Family Agelenidae*

Web—Funnel Web

Webs are around a half foot across, but with a large central hole; the funnel. Though they can be found outdoors under stones, wood piles, rock crevices and woods, we are most likely to see them indoors; often in corners of basements, garages and outbuildings. Indoors, the spider may remain active all year. Spider sits in the hole as a retreat or hiding place. Old abandoned webs indoors may last a long time; often getting coated with dust.

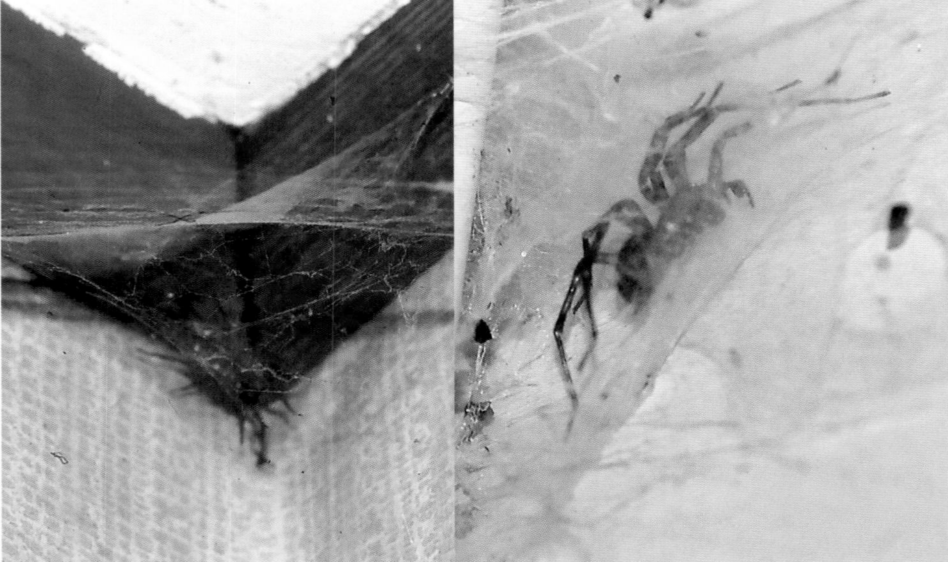

Observations: I have found these spiders and webs indoor often and frequently see them throughout the winter. In the late summer, I have found their webs on the outside of buildings. Some of the webs that I have found were made and used by the males. In outbuildings like barns and garages, webs may last for months.

Barn Funnel Weaver | *Tegenaria domestica*

Spider

Body of female is 7 to 12 mm; male nearly the same size. Legspan 20 to 30 mm. Abdomen is brown, often with chevron markings. Spinnerets are harder to see than some other members of this family. Spider is not native to North America.

Though all these photos were taken indoors, they also can occasionally be found outdoors.

Funnel Webs

Bennett's Hackledmesh Weaver *Family Amaurobiidae*

Web—Funnel Web

Web is small with a large hole in the center which is used by spider as a retreat or hiding place. Egg sac may be placed here too. Irregular loose threads are found around the central hole. Web can be under rocks, boards, bark and logs; not out in the open. Spiders are nocturnal.

Observations: I think every one of these spiders I have seen was discovered because I moved a log, board or rock. Webs are found in such sites and the spider remains here. I have not found their webs indoors.

Bennett's Hackledmesh Weaver *Callobius bennetii*

Spider

A rather robust body of 5 to 9 mm and a legspan of 12 to 18 mm. Carapace and abdomen are about the same size. Both are dark; chevrons are on the thicker abdomen. Spiders are usually seen only when turning over their home sites: logs, boards, rocks and such. I have not seen them indoors.

Orb Webs

Circular webs, usually vertical, ranging from small to large. Since most of the adults die in fall, new and large webs are found mostly in late summer; a few can be seen in spring.

Spiders usually sit in the center (hub) for all or part of the day. Many construct a retreat nearby where they hide when not on the web.

Families of Orb web building spiders:
Araneidae: Orbweavers (most common)
Tetragnathidae: Longjawed Orbweavers
Uloboridae: Cribellate Orbweavers

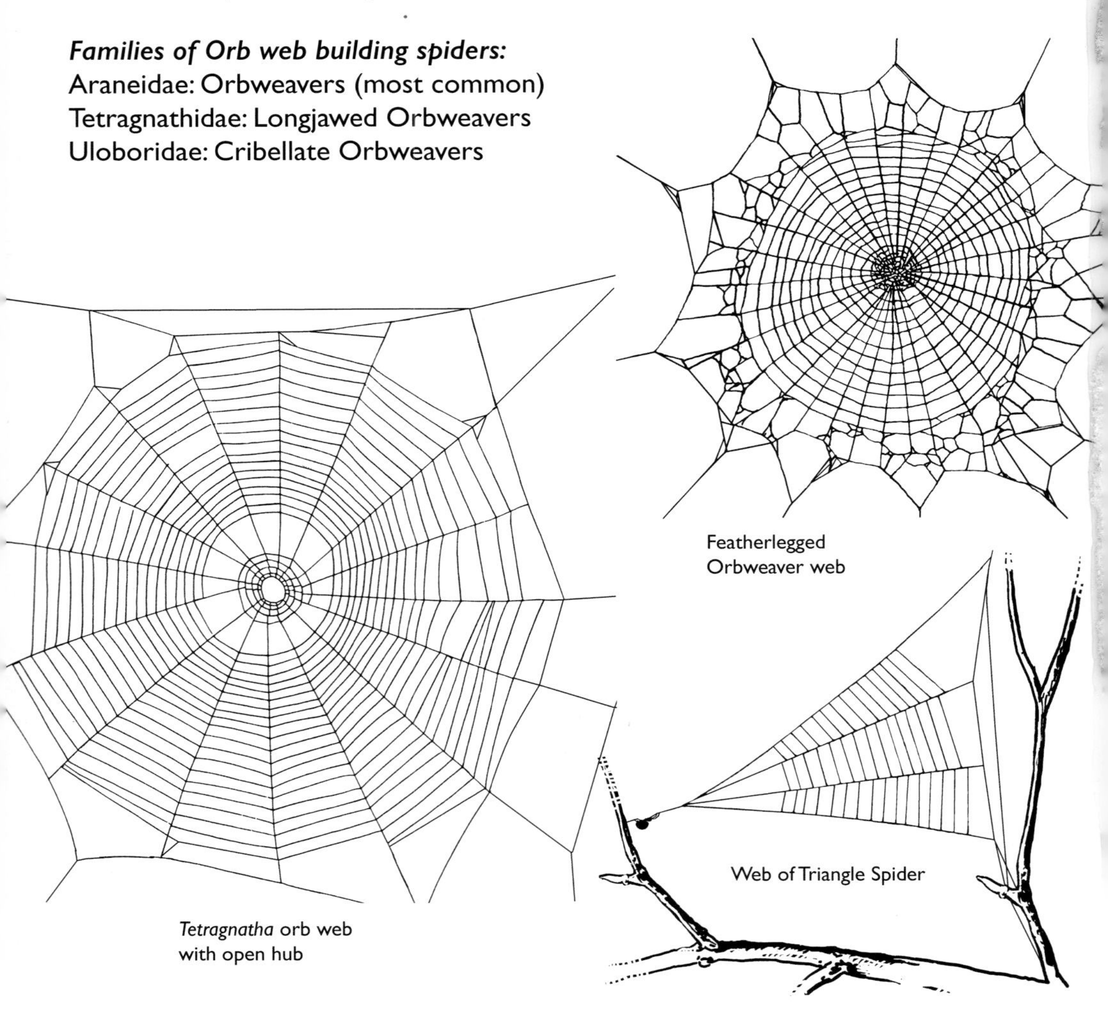

Featherlegged Orbweaver web

Web of Triangle Spider

Tetragnatha orb web with open hub

Orb Webs

Orb Webs

Star-bellied Orbweaver *Family Araneidae*

Web—Orb Web

Web is medium-sized, 6 to 10 inches, but with the framework, it may look larger. Webs are found in fields of tall grasses or nearby low bushes; often within 4 feet of the ground. Webs are vertical with 20 to 25 radii and an open hub. Webs are made at dusk and spiders are present in the web at night, but may still be in the web at dawn.

Observations: A walk in a grassy field on a dew-covered morning in late summer reveals many orb webs. This spider species is not the most common. Often the dew-covered spider is still in the hub or over at the edge in a small retreat. I have not seen their webs in the woods at this time.

Star-bellied Orbweaver *Acanthepeira stellata*

Spider

Body is 10 to 12 mm with a legspan of 12 to 15 mm. The rounded brown abdomen is distinct with 12 cone-shaped projections; 4 or 5 on each side with others at the front and rear.

Giant Lichen Orbweaver | *Family Araneidae*

Web—Orb Web

With the large spider, the vertical webs are also huge, maybe three to four feet in diameter with 15 to 20 radii and a closed hub. The webs are made at night or at dusk. They are hung in the forest between trees. Spider is in the hub at night, but in the day goes to a retreat on nearby trees. This retreat could be among lichens on the trunk, where the spider's colors blends in with the lichens. Webs are actively used in summer; a warm weather spider.

Observations: Finding the web of this spider is a delight. The huge snare may be placed right on a trail and you can't help but notice it. A recent one that I found on a trail was remarkable. The catching spiral portion was about three feet in diameter, but the initial frame constructed to build the web was ten feet long! It went from a tree on one side of the trail to a tree on the other. I came back to the web often over the next couple of weeks. When not in the hub, the spider was among lichens on a nearby tree.

Giant Lichen Orbweaver *Araneus bicentenarius*

Spider

With a huge abdomen, the female's body is 20 to 30 mm and a legspan that may reach 50 mm. The mostly rounded abdomen is green with two zigzag markings on the top. Color and pattern allows them to be well camouflaged on lichen-covered tree trunks. Two lateral humps are towards the head end. Carapace is small and dark.

undersides

Orb Webs

Cross Orbweaver | *Family Araneidae*

Web—Orb Web

Webs are vertical, large, two to three feet in diameter with about 30 to 35 radii. Webs are in gardens, yards, tall grasses, shrubs, small tree branches and eaves of buildings. Webs are often used all day with the spider in a retreat made of a rolled up leaf; in the hub at night. Webs are active in summer and into fall.

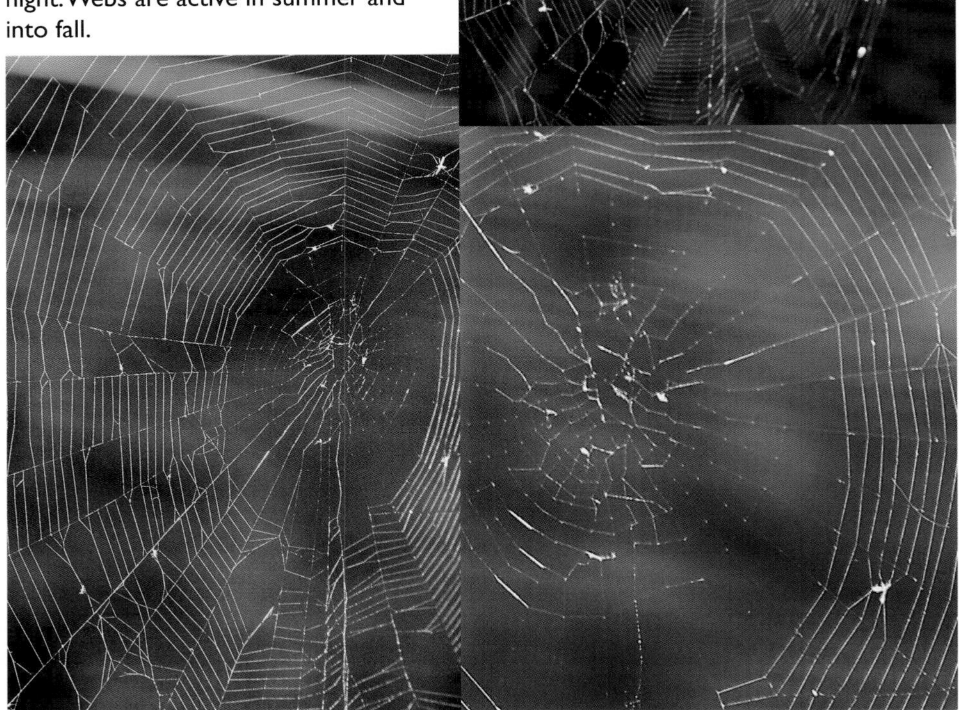

Observations: This a common spider found frequently around our homes. I have found their webs regularly in the late summer on the garage, under the eaves of the roof. An interesting web that I located recently was made on a clothes line. The upper frame was attached to the clothes line while an anchor thread went all the way to the ground, about six feet. The spider made a retreat out of a knot that I tied there.

Cross Orbweaver — *Araneus diadematus*

Spider

Very large circular abdomen; the body is 10 to 20 mm with a legspan up to 40 mm. Abdomen is brown with zigzags on the back. Towards the head end, there are two prominent humps and white spots in a cross shape. This non-native spider is also called the "garden spider" by some.

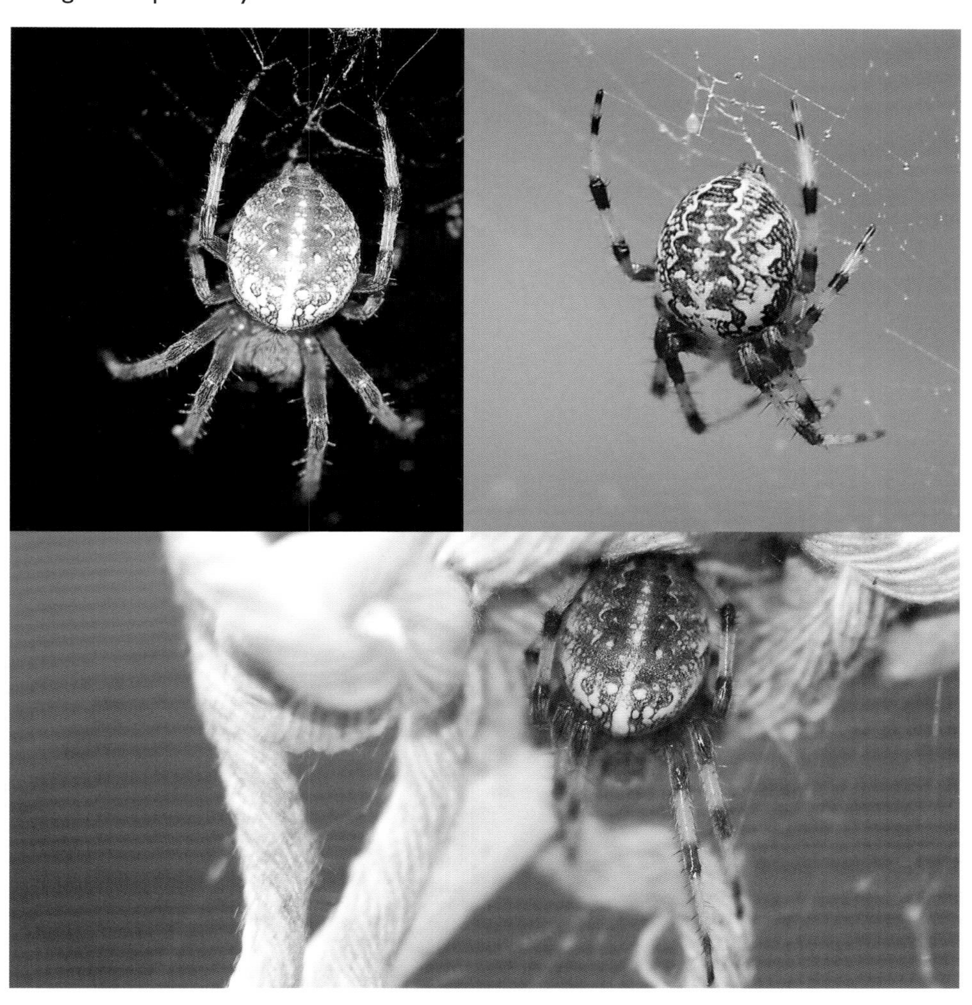

Catfaced Spider | *Family Araneidae*

Web—Orb Web

A large-bodied spider that makes a large web. Web with about 20 radii and a closed hub. Web made in the evening for night hunting. Spider in retreat during the daytime. Webs made in barns, under bridges and cliffs. Webs in late summer.

Observations: Sometimes called a "barn spider," as webs are mostly found in such sites. I have found this spider and web in late summer.

Catfaced Spider *Araneus gemmoides*

Spider

A large body of 13 to 25 mm and a legspan of about twice that. Abdomen is brown with two pronounced humps towards the head end. The name of cat-faced refers to the light markings on the front end of the abdomen near the carapace (bottom photos). Spider is a sedentary hunter and the abdomen will become large and round by the time it lays eggs.

Orb Webs

Marbled Orbweaver — *Family Araneidae*

Web—Orb Web

Web is vertical, large, 20 to 30 inches in diameter with 20 to 30 radii, a closed hub and a curled leaf for a retreat. Webs are made in meadows, grasses, gardens, shrubs, and undereaves of buildings; from 3 to 10 feet above the ground. Webs made at dusk and spider sits in the hub at night; retreat daytime.

Observations: I have found these large webs frequently in late summer in grasslands, shrubs and on buildings. Easiest to find these webs at dusk on summer days; will last into the fall. A recent web that I found was on a road sign. The spider went between the sign and the sign post to hide in its retreat. It was very hard to see in the daytime.

Marbled Orbweaver *Araneus marmoreus*

Spider

Body is from 9 to 18 mm in length with a legspan of about twice this. Abdomen is rounded without bumps. Variable in color; from brown to gray to orange all with a scattering of spots; even one phase with a thick black markings.

Orb Webs

Nordmann's Orbweaver — *Family Araneidae*

Web—Orb Web

The large vertical web of about two feet across has 25 to 30 radii, a closed hub and a retreat of a curled leaf if the web is in a tree. Webs are also in shrubs, tall grasses, and the sides of buildings. Webs are made at dusk and the spider sits in the hub at night. A new web is made each evening.

Observations: I have found this spider web on the side of the garage regularly. Webs are best seen in late summer. Spider often in a retreat or hiding in cracks in the daytime and hard to see, but quite easy to find after dusk.

Nordmann's Orbweaver *Araneus nordmanni*

Spider

Body is 10 to 20 mm with a legspan of up to 40 mm. A gray oblong abdomen with slight bumps towards the front, a zigzag pattern in the center with dark coloring; sometimes black between the zigzags.

Note the variability of the abdomen markings of this species.

Orb Webs

Shamrock Orbweaver | *Family Araneidae*

Web—Orb Web

The large vertical orb web is 20 to 30 inches in diameter. Webs have 15 to 25 radii with a closed hub. Webs are made new each evening; spider in hub at night. A retreat made from curled leaves or other parts of a plant is where the spider is in the daytime. Webs are very common in fields, gardens, meadows and bushes. I find them abundant among the asters and goldenrods of late summer and early fall. I have never seen their webs in the woods, though they may be found at the edge of buildings. Several threads are attached to the retreat.

Observations: These large orb webs are abundant in fields and gardens. Though the webs are made at dusk, I find the spiders quite easily in the early mornings while they are still in the hub of the web; often covered with dew; some are in the retreats.

Shamrock Orbweaver *Araneus trifolium*

Spider

Large spider from 10 to 20 mm and a legspan of up to 40 mm. The big round abdomen is variable in color but always with a spotted pattern; this pattern can vaguely resemble a shamrock and hence the common name. Huge abdomen, very small carapace and tiny eyes are distinctive. I am probably asked about this spider more than any other.

Shamrock Orbweaver webs

Shamrock Orbweaver webs

Orb Webs

Openfield Orbweaver | *Family Araneidae*

Web—Orb Web

Smallest of the *Araneus* orb webs; less than one foot in diameter. Webs are usually vertical, but may be inclined or even horizontal. Webs have 15 to 25 radii and a closed hub. They are found in tall grasses, fields and shrubs. Spider in the web at night. Webs are found in the late summer into fall.

Observations: I find these small orb webs in the fields of late summer and early fall. Being less than a foot in diameter, they can get over looked amongst the much larger and more obvious orbs. Most are vertical, but I've seen inclined webs and even a few horizontal. Webs are made at dusk with the spider present at night, but I have found many still in the dew-covered webs in early morning.

Orb Webs

Openfield Orbweaver *Araneus pratensis*

Spider

Body is only about 5 mm long with a legspan of about 10 mm. The oval abdomen light with dark stripes or an orange color with light lines. Some that I have seen have only a dark spot towards the end of the abdomen.

Six-spotted Orbweaver — *Family Araneidae*

Web—Orb Web

Web is small, only six inches in diameter. Webs are placed in shrubs or broad-leaf trees about four to six feet above ground. Webs may be in grasslands as well, but still under broad leaves, such as milkweed. They may be in these sites from spring into summer. Spider sits in the hub during the day; no retreat.

Observations: I find these small orbweavers in late summer in fields, but never many. Some of the best observations I have had of this spider have been in their small web under the curled leaf of milkweed plants. The egg sac may be in the web too.

Six-spotted Orbweaver *Araniella displicata*

Spider

A body of 5 to 9 mm with a legspan of about twice that. The oval abdomen may be white, yellow or pinkish with the characteristic three pair of black spots.

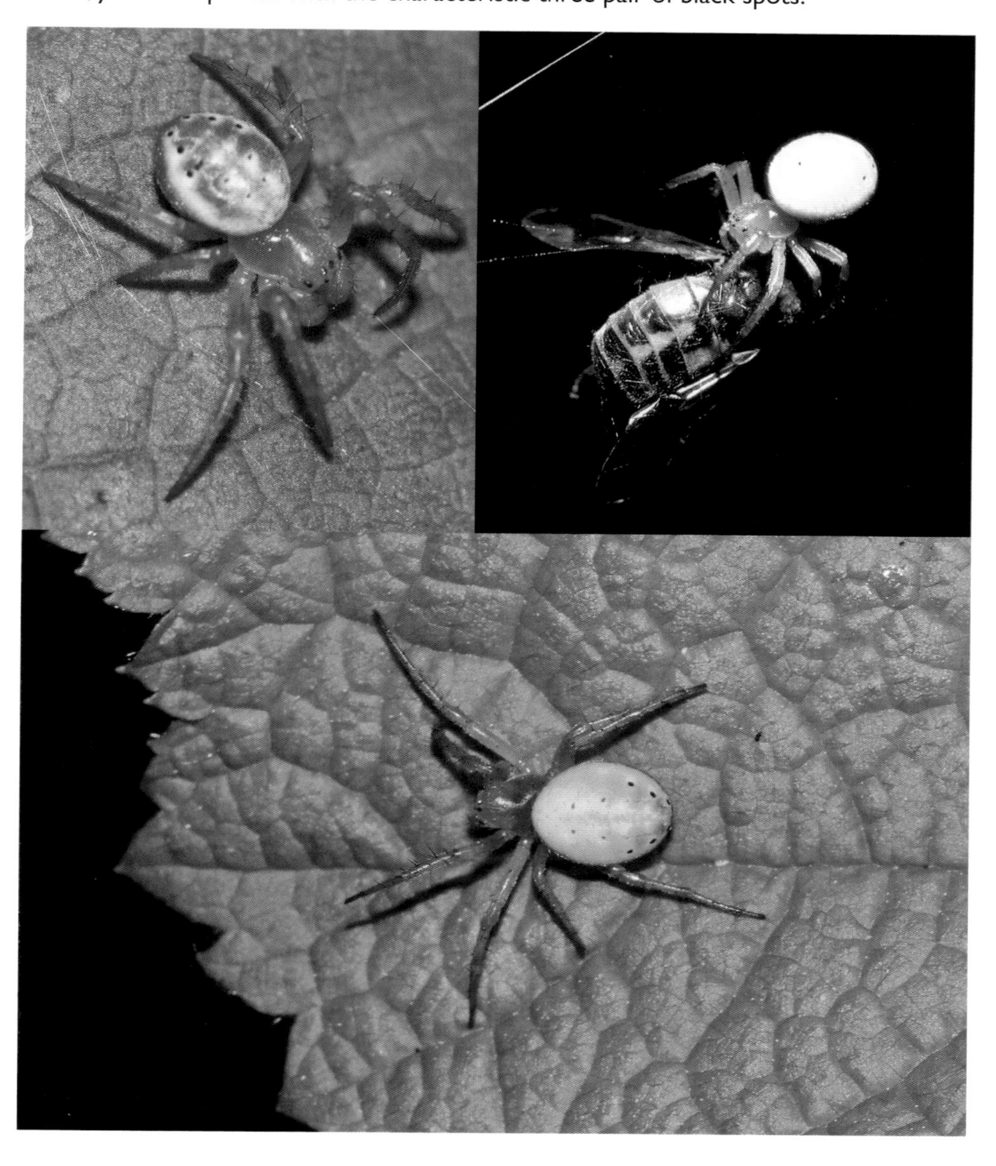

Orb Webs

Yellow Garden Spider | *Family Araneidae*

Web—Orb Web

Web is very large, often three feet or more in diameter. Hub is tightly woven, usually with a zigzag stabilimentum that extends vertically several inches to nearly a foot. 30 to 50 radii and no retreat. Webs are usually in sunny sites of meadows, fields, gardens and yards. Best seen in late summer into fall. Spiders are in the hub in daytime.

Observations: This spider is not so common in the North Woods. It is more common to the south of here, but in recent years I have found their webs. Webs are always in open sites. I have never seen their web in a tree, so most are not far above the ground. Spider sits inverted in the stabilimentum in the daytime.

Yellow Garden Spider *Argiope aurantia*

Spider

Spiders are large; 20 to 30 mm with a legspan that could extend to 70 mm. The ovoid abdomen is mostly yellow with a large black band with yellow markings in the center. Spiders sit inverted in the center of the web, holding their two front and two rear pairs of legs together; looking almost like the spider has only four legs.

Banded Garden Spider | *Family Araneidae*

Web—Orb Web

The large orb web is about two feet in diameter. Found in grassy meadows among grasses and wildflowers. Tightly woven hub, often with stabilimentum, 30 to 50 radii and no retreat. A few of these webs may have a veil-like web over the main orb web. This seems to protect the spider better. Webs are common in late summer into fall.

Observations: A walk on a dewy morning in late August or early September never fails to reveal at least one of these webs; often with the spider "at home." They are diurnal, but I usually see them in morning and since they are coated with dew, they may have been in the web during the early hours. Like other orbweavers they have poor eyesight, but with excellent senses, they often detect me and drop down before I get close. I have witnessed mornings in early September of more than a hundred of these webs in a meadow, but with the arriveal of cool weather two weeks later, I could find none.

Banded Garden Spider *Argiope trifasciata*

Spider

Large; 15 to 20 mm and a legspan that could reach 60 mm. Bands of yellow, white and black on the oval abdomen; carapace is light colored. Sitting in the web inverted, the two front and two rear pairs of legs are held together; looking almost like the spider has only four legs.

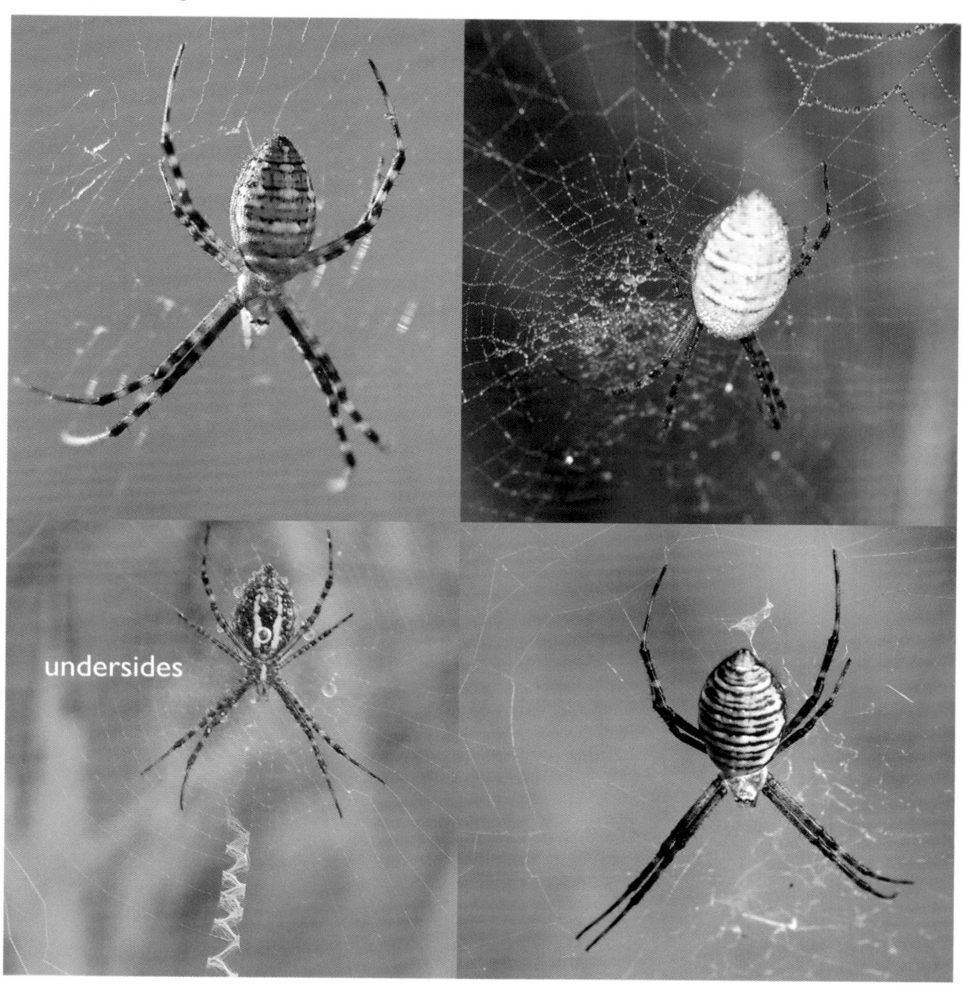

Orb Webs

Conical Trashline Orbweaver *Family Araneidae*

Web—Orb Web

These small spiders make small webs of about a 12 inch diameter. Spirals may extend to or near the hub of the web. Many spokes; 30 – 50 with no retreat. The "trashline" name refers to the spider's habit of placing debris and captured prey in the central line to help make a stabilimentum. Webs are in the woods, about five to six feet above the ground; spring to summer. The spider is in the center day and night.

Observations: I think all of the trashline orbweaver webs I have found have been in the woods in the summer; most in deciduous forests. Webs may be hard to see, but I have learned to walk into the sun, using backlight to see them best. The spiral threads going all the way to the center appears to be unusual. These orbwevers can be hard to see as they sit amidst their trashline providing excellent camouflage for the tiny spider.

Conical Trashline Orbweaver *Cyclosa conica*

Spider

Spiders are small; 5 to 7 mm with a legspan about twice this. The name *conica* refers to the cone shape of the abdomen. Abdomens are light brown and when in the web, the spider may be hard to see.

The bottom photo shows a blood-filled mosquito wrapped up by a Trashline Orbweaver.

Humped Trashline Orbweaver *Family Araneidae*

Web—Orb Web

Webs are about a foot in diameter and about five or six feet from the ground. Found in deciduous or coniferous woods but has been known to inhabit fields, gardens and yards. The small spider makes a web with 30 to 50 radii. A line of debris in the center of the web forms a type of stabilimentum. These small spiders may be hard to see among all the debris in the center of the closed hub. Spiders remain in the center of the web; no retreat.

Observations: I've only seen this spider and its web in forests, but several of the photos above were taken by a photographer on a backyard playset. The line of debris in the center of the web is distinctive of the two species of trashline orbweavers.

Humped Trashline Orbweaver | *Cyclosa turbinata*

Spider

Small; only 5 to 7 mm with a legspan of maybe 10 mm. The abdomen is light colored, pointed like a cone, but on the front end are two pronounced humps.

Humpedbacked Orbweaver *Family Araneidae*

Web—Orb Web

A small to medium-sized web; maybe12 to 18 inches across with 15 to 25 radii and an open hub. Spider does not make a retreat, but frequently sits on nearby branches and twigs when not in the web. The web is in trees and shrubs in woods, sometimes near water. Webs are made at night and taken down in the morning. Overwintering in the penultimate stage, they are already making webs in spring; spiders continue making webs through the summer and into fall.

Observations: I typically see these webs in the evening, being made shortly after dusk. Early the next morning, in proper light, I see more of these webs. I think every one that I have found has been in a deciduous forest among the lower branches of trees. The open hub is a good feature to look for. If the spider is not in the hub, it may be on twigs and branches nearby; appearing almost like a tree bud.

Humpedbacked Orbweaver *Eustala anastera*

Spider

Spiders are small; 6 to 10 mm with a legspan of 10 to 20 mm. Abdomen is gray with a hump or point extending on the rear end. Zigzag markings on the abdomen. I frequently see this spider in its web in late summer or fall.

Spinybacked Orbweaver *Family Araneidae*

Web—Orb Web

The web is supported by foundation lines on lower tree branches and bushes from 3 to 6 feet above the ground in open woods. The web, which can be three feet in diameter, consists of 20 to 25 radii, a tightly spaced spiral, and no spiral near the center of the hub. The spider hangs inverted in the center of the web. The bridge threads on which the rest of the web is suspended may be very long; more than ten feet. This makes the web look larger than it actually is.

Observations: Spinybacked Orbweavers are often conspicuous in the center of their webs. I am impressed by the numerous radii and closely-spaced spirals of the 3-foot diameter web, but it is the length of the bridge threads that can be more than ten feet long, which makes this web one to remember.

Spinybacked Orbweaver *Gasteracantha cancriformis*

Spider

The half-inch (13 mm) female is one of the most distinctive of American spiders. The abdomen has a hard exoskeleton with six spines around the edge. This abdomen, with its lateral spines, is wider than long. Colors on the abdomen vary from white to yellow to orange-red with the spines being black or red. A couple rows of black spots dot the top of the abdomen. Despite the various colors on the abdomen, the cephalothorax is always black. Legs are short on these sedentary hunters. Males have a black-white abdomen and are about one-fourth her size. Spinybacked Orbweavers are common in the southeast U.S. ranging from North Carolina to Florida and extending west to California.

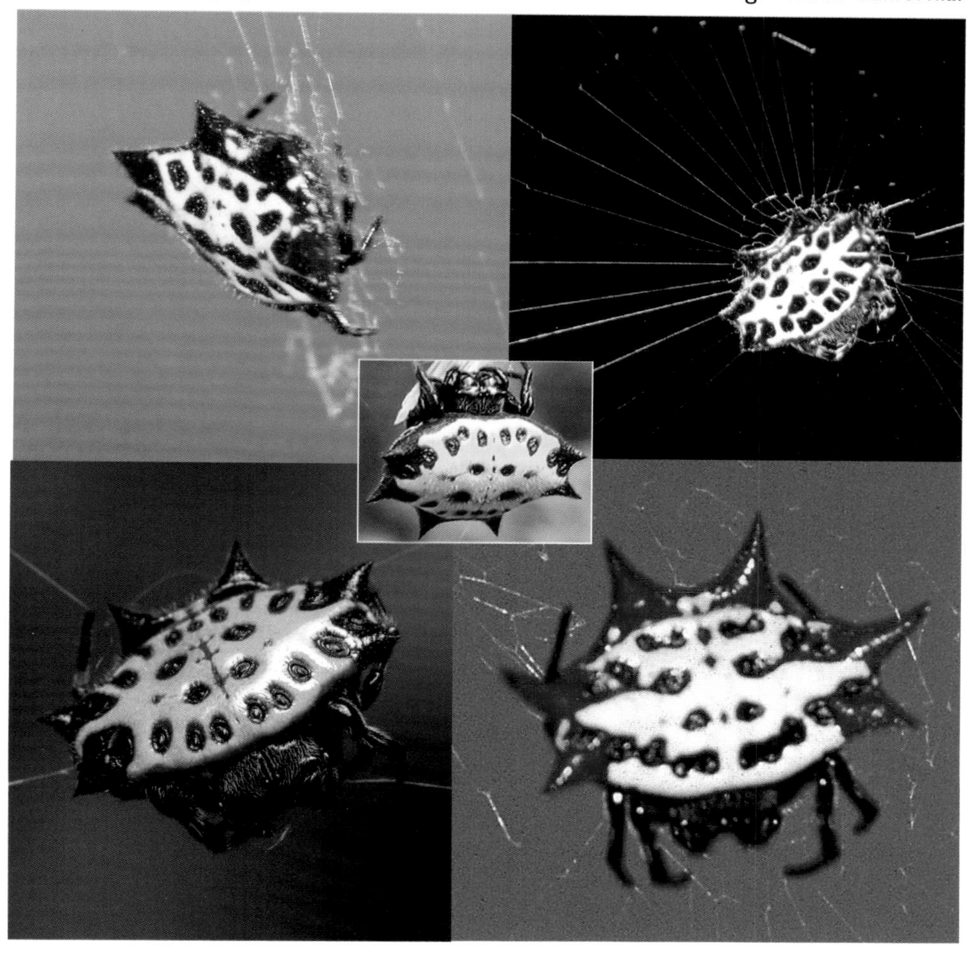

Furrow Orbweaver *Family Araneidae*

Web—Orb Web

Webs are medium to large. Though usually about two feet in diameter, they are sometimes larger; 15 to 20 radii, widely spaced spirals, closed hub. Webs are made on a variety of places: walls, fences, tree trunks, bridges, docks, shrubs, sides of buildings, on branches near water and even in grassy fields. Webs are made in the evening; used at night. One of the earliest orb webs each spring. Overwinter as subadults.

Observations: Maybe the most common orb web that I see throughout the whole season. Orbs are made in spring; I have seen them as early as April. The only orb web that I have ever seen entirely covered with snow. Common as aerial webs in a swamp that I visit regularly (note page 122 upper right). Webs have a long season and are built on a variety of sites. Spiders may make their own retreat out of silken material.

Furrow Orbweaver *Larinioides cornutus*

Spider

Medium-sized; the body is 6 to 14 mm with a legspan of 18 to 35 mm. The light brown oval abdomen has a dark center outlined with a zigzag pattern. Within this dark part is the light-colored cross pattern. The undersides of this species reveals a yellow parenthesis pattern as it sits inverted in the center of the web (not the only spider species to show this mark).

Lined Orbweaver | *Family Araneidae*

Web—Orb Web

A small web is made by this tiny spider. Web is about one foot in diameter, but with many radii; usually 50 to 60. Web is mostly vertical, but slightly inclined. Spider sits in the closed hub. Webs are constructed in several sites including roadsides, edges of deciduous woods and tall grasses. Spiders are diurnal and may be in their webs well into fall. Some form a "bull's eye" stabilimentum in the center. Retreat is in curled leaf.

Observations: Despite their tiny size, I have seen these webs late into fall. Distinctive of a *Mangora* web are the many radii. Best to have the light in front of you to more easily find and see these small webs.

Lined Orbweaver *Mangora gibberosa*

Spider

Tiny; 4 to 6 mm with a legspan of about 5 to 10 mm. The light-colored oval abdomen has stripes and lines on it.

Note the circular stabilimentum in the bottom photo.

Orb Webs

Tuftlegged Orbweaver | *Family Araneidae*

Web—Orb Web

Web is small, only about a foot in diameter, but with 50 to 60 radii (30 to 70) and many spirals. Web is vertical with a slight incline. Spiders sit in the closed hub. Webs are mostly in deciduous forests at about five feet from the ground. Webs may continue to be active far into the fall.

Observations: I always find these webs in deciduous forests. They are tiny with many radii. And despite their size, they remain active well into October. Spiders are nearly always in the center of the web.

Tuftlegged Orbweaver *Mangora placida*

Spider

Tiny; only 2 to 5 mm. Legspan is 5 to 10 mm. The oval abdomen is light along the sides with a dark band in the center. The adult of this spider is often smaller than immatures of other spiders.

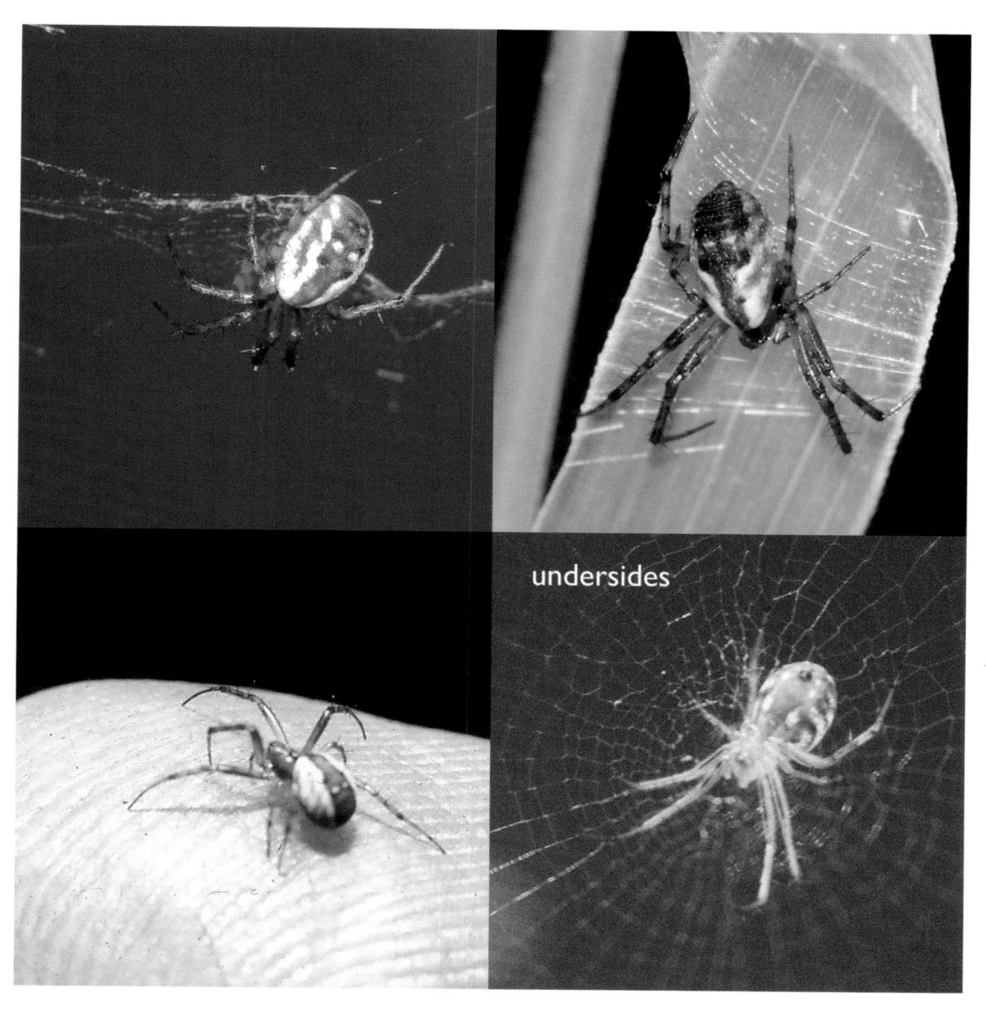

Basilica Orbweaver | *Family Araneidae*

Web—Orb Web

One of the most unique orb webs, and its form gives this spider its common name. They construct an orb web in the horizontal position that has many radii very closely spaced. After the web has been spun, the spider attaches a thread to the hub and pulls it up forming an uplifted center. It is this shape that reminded someone of the dome of a cathedral and led to this spider's common name of "Basilica Orbweaver." Spiders hang under the dome. While most orbweavers rest in the web in an inverted vertical pose, this spider is inverted horizontally; more like what we might see in the sheetweb weavers (Linyphiidae). Webs are made in shady areas within or under shrubs or forest understory.

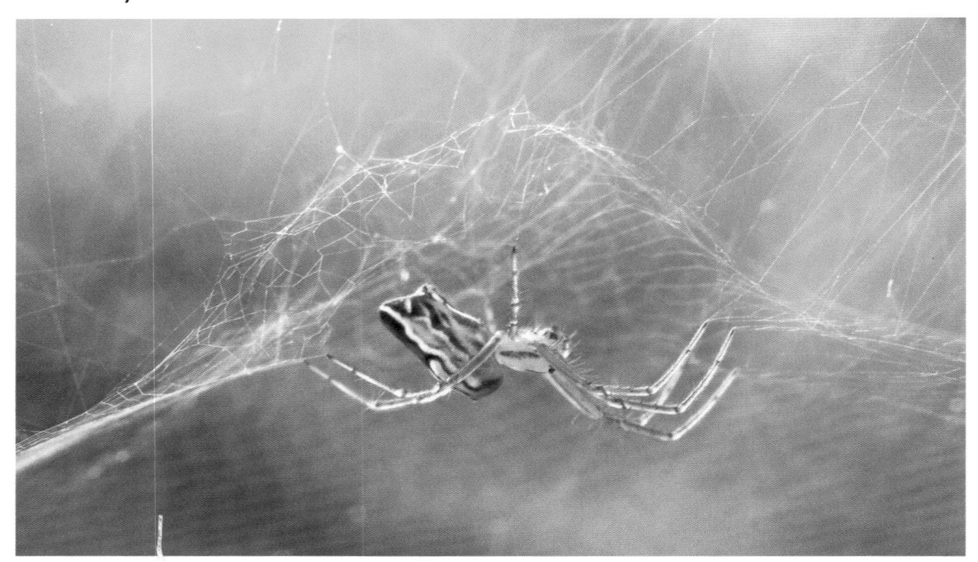

Observations: Though the web is orb and the spider is classified as a member of the family Araneidae, this domed "basilica" web can cause it to be confused with the domed web of the Filmy Dome Spider (*Neriene radiata*). And it does superficially resemble this web, but a closer look reveals radial and spiral threads. Also the spider makes a network of irregular threads; making the web to look a bit like that of the Labyrinth Spider (*Metepeira labyrinthea*). The orb web of the Labyrinth Spider is vertical while the Basilica Orbweaver's web is horizontal.

Basilica Orbweaver *Mecynogea lemniscata*

Spider

The length of the female's body is less than one-half inch (12 mm). With outstretched legs, it can reach to about one and a half inches (38 mm). The long abdomen is yellow-green and orange; with olive-green patterns on the side; underside is dark. The cephalothorax is yellow to light brown; a dark line in the middle of the dorsum.

Labyrinth Spider | *Family Araneidae*

Web—Orb Web

Web is of two parts— The vertical orb of small to medium size and an irregular labyrinth of threads, looking almost like a web of Theridiidae; slightly above and behind the orb web. Unique retreat is cone-shaped and adorned with debris and prey parts (photos this page).

Observations: Spiders are widespread across the United States but more common to the south of the North Woods.

Labyrinth Spider | *Metepeira labyrinthea*

Spider

5 to 6 mm long with a legspan of 7 to 15 mm. Oval abdomen of light brown has a white arrowhead shape in the center.

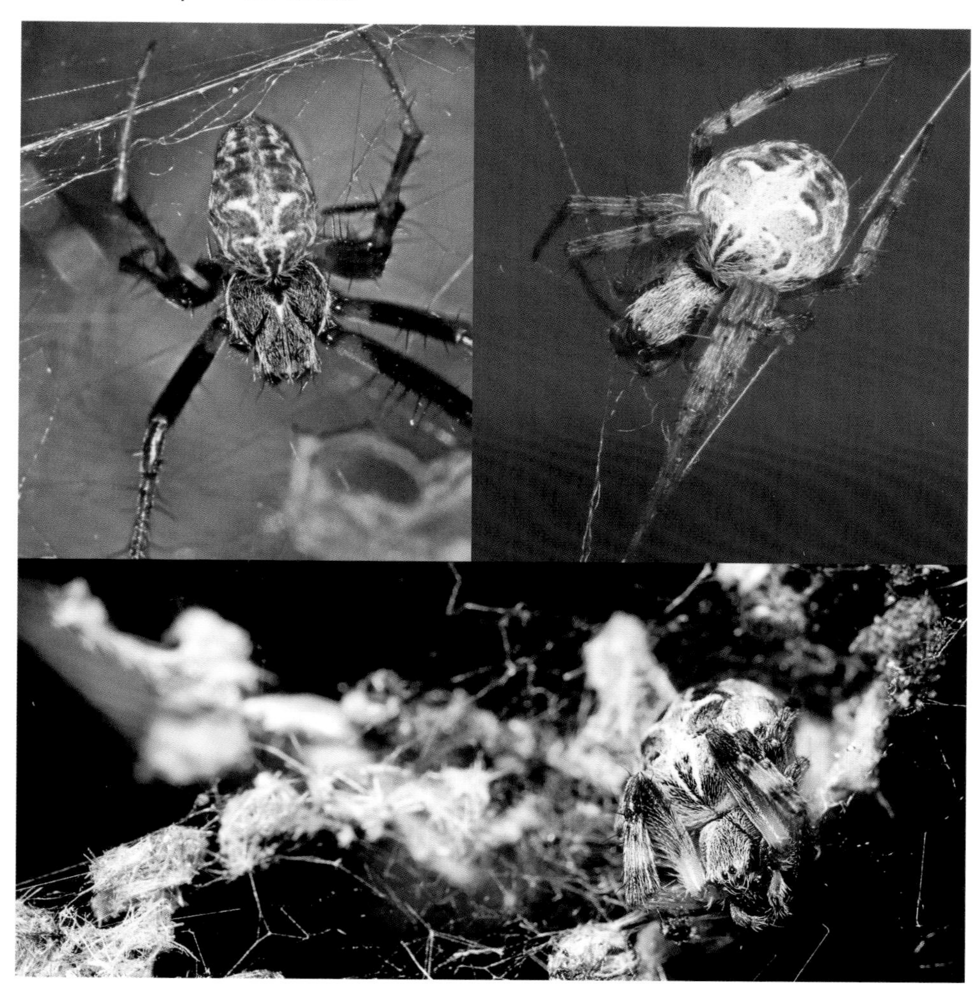

Spined Micrathena | *Family Araneidae*

Web—Orb Web

The orb webs are medium-sized; ten to twelve inches in diameter with 35 to 45 radii; hub is open. Webs are in deciduous forests in the shade, often encountered along wooded trails. Constructed three to seven feet above the ground, the spider needs to have long frame threads giving the appearance of a much bigger web. Spider sits in the center day and night.

Observations: This member of the *Micrathena* genus is found further south than others. I have often seen them a little to the south of the North Woods. Spider with its spined body sits in the open hub of the web and since webs are often over or near trails, we may encounter the spider before noticing the web. Most of the times that I have found this spider in its web, it has been on a hot summer day.

Spined Micrathena *Micrathena gracilis*

Spider

Body is 7 to 10 mm long and since they are highly sedentary, the legspan is only slightly more. Abdomen appears to be too large for such a small spider. Abdomen is mostly light with five pair of spines, mostly along the sides. The smaller carapace is dark; legs are short.

White Micrathena | *Family Araneidae*

Web—Orb Web

The orb webs are small, from 6 to 8 inches with 35 to 40 radii. Hub is open and the web is made among the trees of deciduous forests. Spider sits in the center of the web day and night in the shady woods. Since the web is constructed between trees, sometimes over trails, the frame webs may be long, making the web appear much larger than it really is. Web is 3 to 7 feet off the ground.

Observations: I have found this web with the spider in the center while walking deciduous forest trails on hot summer days. This is the smallest member of the *Micrathena* genus and a hiker can walk right into the web without seeing the spider.

White Micrathena | *Micrathena mitrata*

Spider

Body is only 4 to 5 mm long. The mostly white abdomen is at least half of the length. Short legs so legspan not much more than body length. Abdomen has two pairs of small spines towards the rear end. Carapace is dark.

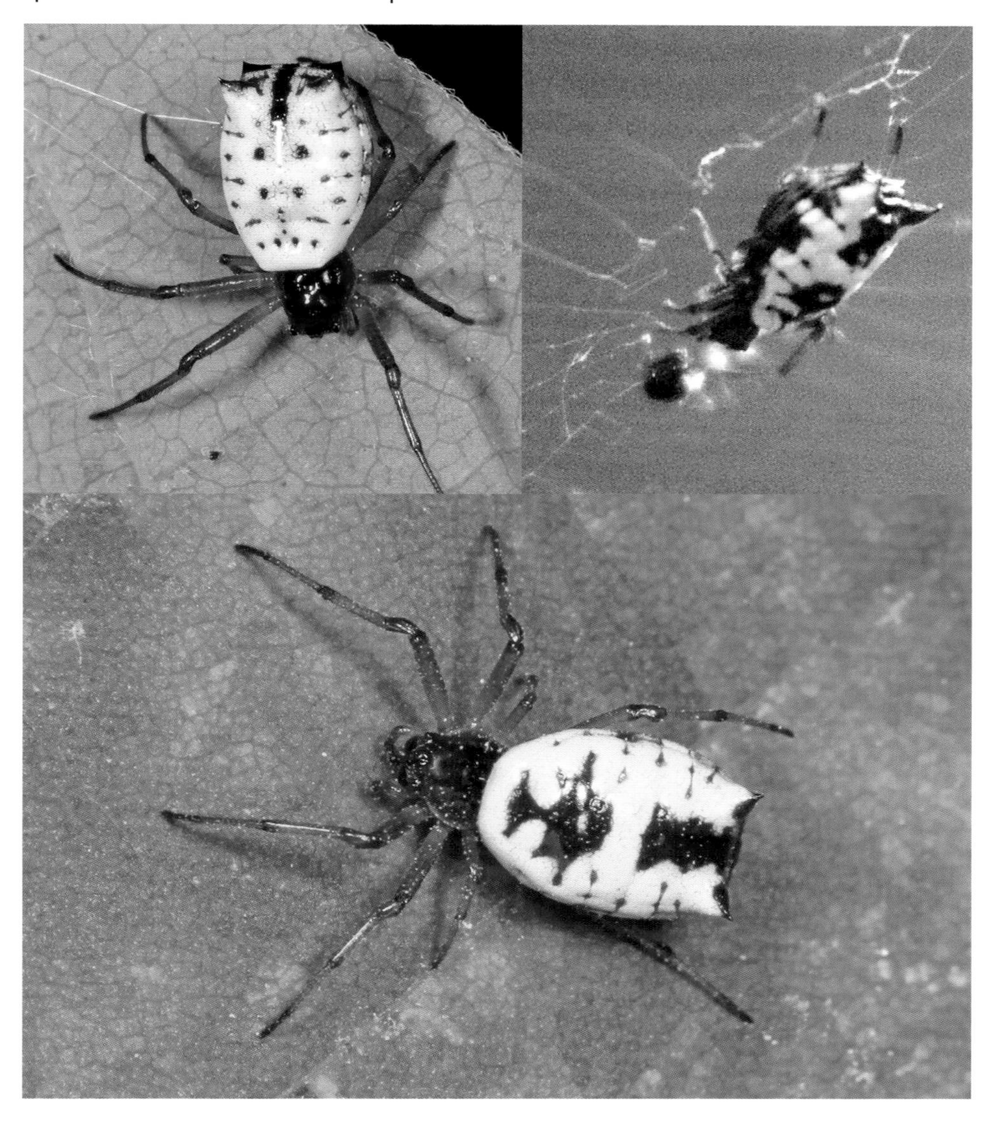

Arrowshaped Micrathena *Family Araneidae*

Web—Orb Web

The orb web is about a foot in diameter with 30 to 40 radii. Hub is open, often with a short stabilimentum. Webs are vertical, but may be reclined towards horizontal. Webs are usually in open woods, gardens and yards about three to five feet above the ground. Spider sits in center day and night in summer.

Observations: Though the webs are usually vertical in and among trees, one that I found was nearly horizontal and on the surface of an air conditioning unit just outside of a building. Despite its uniqueness, it had a short stabilimentum around the hub.

Arrowshaped Micrathena | *Micrathena sagittata*

Spider

Largest of the *Micrathena* genus, Arrowshaped have a body length of 8 to 10 mm and longer legs; giving a legspan of about twice the body length. Abdomen is yellow and red with three pair of spines; the ones in the rear are the longest, giving the appearance of the shape of an arrow. Carapace is reddish-purple.

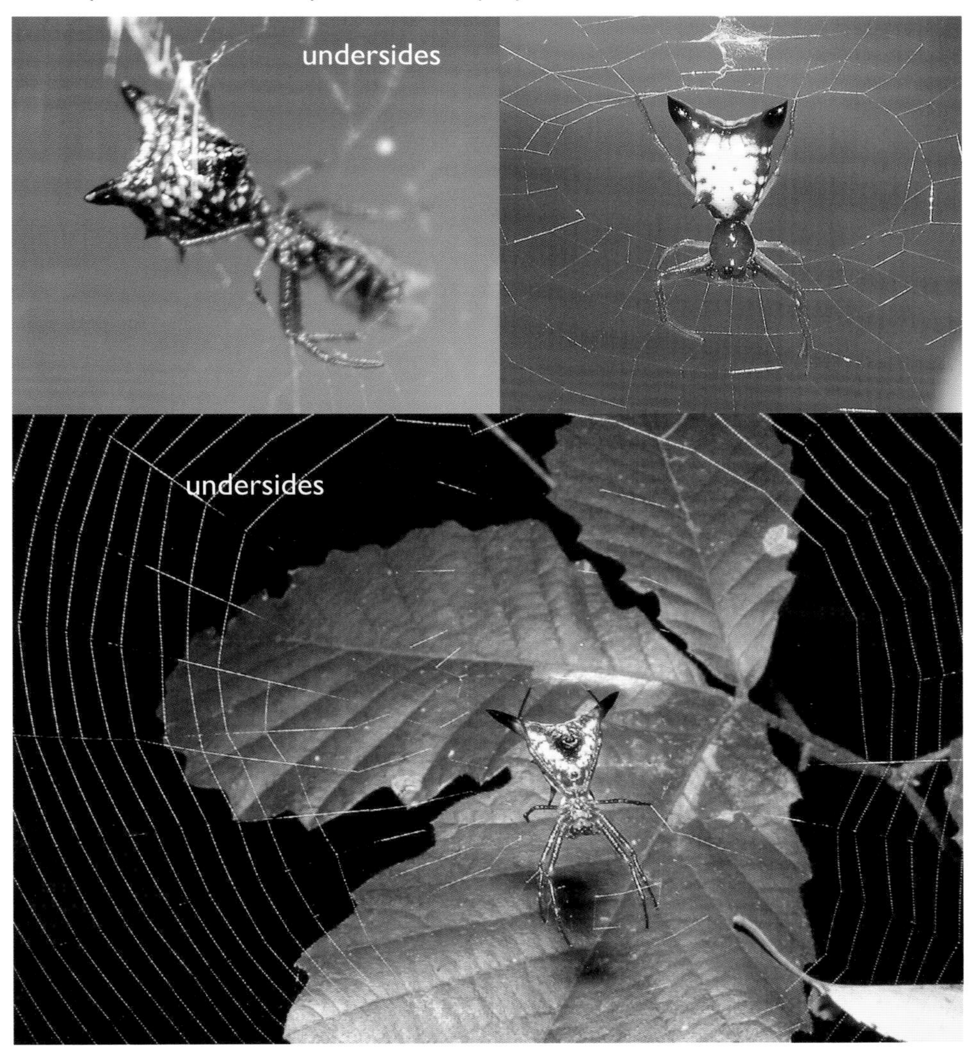

Orb Webs

Arabesque Orbweaver — *Family Araneidae*

Web—Orb Web

The medium-sized orb web may be from 6 to 18 inches in diameter with about 20 radii. Webs are made in sunlit open sites in tall grasses and low bushes of fields and woods edges. The spider sits in the open hub with its abdomen sticking through. Webs are occupied both day and night in summer; usually no retreat.

Western Spotted Orbweaver
Neoscona oaxacensis

Observations: Once I acquired the eye for it, I found the webs of this spider to be common in fields. Though the spider may be in the web all night, I have frequently found them on dewy early mornings. Not making a retreat, they do go to the sides of the web as a shelter and hiding place. The open hub is a great aid in identification since most field spiders do not have an open hub in their webs.

Arabesque Orbweaver *Neoscona arabesca*

Spider

Body of 5 to 12 mm with a legspan of 18 to 30 mm. The oval abdomen is light brown with white angular markings on the center. To the side of this white pattern are five pairs of black dash markings.

Western Spotted Orbweaver *Neoscona oaxacensis*

Orb Webs

Golden Silk Orbweaver — *Family Araneidae* (**sometimes in Nephilidae*)

Web—Orb Web

Anyone who has seen this web would note its uniqueness. The web is an orb that is huge; maybe up to three feet in diameter. Snares are placed in shady sites at least three feet off the ground. Threads are very strong. The many non-sticky radii are white; the viscid spiral is golden-yellow (this explains the common name). Radii are usually not in a straight line; as they are pulled out of alignment and appear to be notched or looped. The large female *Nephila* is diurnal and hangs on the lower side of the inclined web all day long. Often the tiny male, usually no more than one-fourth the size of the female, is also in the web. And the web may also attract small "kleptoparasitic" spiders.

note male at top and "golden" silk of web

Observations: My own experience with this spider and its web has been one of awe and wonder. Not only was I impressed with the size and strength of the web, I was also amazed by the huge spider sitting inverted in the web during daylight hours. A little searching turned up the smaller male in the snare as well. Accidentally walking face-first into this web is a memorable experience.

Golden Silk Orbweaver *Nephila clavipes*

Spider

This is one of the largest orbweavers in the country. The female's body may be 1-inch long (25 mm) and have a leg span of up to 4 inches (100 mm). The cephalothorax is black, but often covered with silver hairs. The abdomen is golden-olive to yellowish-brown with five or six pairs of yellow to white spots on the dorsal side. Conspicuous tufts of hair on the femora and tibia of legs I, II and IV. Spiders are mostly found in the southeast part of the country along the Atlantic and Gulf Coasts.

Orb Webs

Orchard Spider

Family Tetragnathidae

Web—Orb Web

The 12 inch web is reclined to nearly horizontal with about 30 radii and an open hub as often seen in this family. In deciduous woods, sometimes in orchards and shrubs not far from the ground. Webs are occupied in summer with the spider under the tilted web; quick to drop down.

Observations: I have found this spider on summer days as it was sitting the center of its mostly horizontal web. It often has its abdomen tip centered in the open hub. Best to use binoculars to observe it, as I find they are quick to drop off the web. Probably the most arboreal species of this family of spiders.

Orchard Spider | *Leucauge venusta*

Spider

Body about 5 to 10 mm, but with longer legs than seen in most orbweavers, it has a legspan of up to 22 mm. The long ovoid abdomen is silver-white with black markings above and some yellow towards the rear end. Undersides green with some red spots. Fairly long greenish legs.

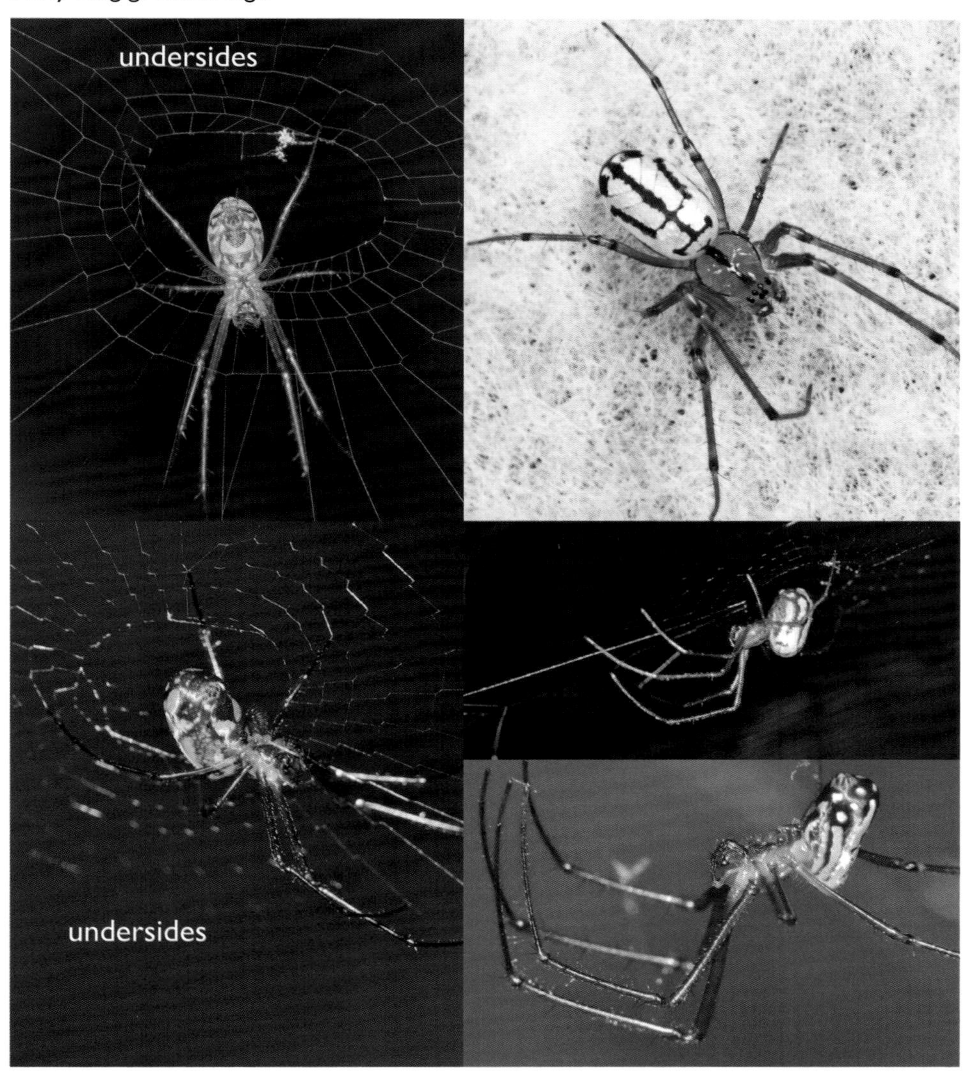

Cave Orbweaver — *Family Tetragnathidae*

Web—Orb Web

The 10 to 16 inch orb web with 15 to 30 radii is reclined towards horizontal, but more vertical than many of the webs of this family of spiders. Webs found on cliffs, cave openings and cellars. Could also be found in culverts or under bridges. Webs may be occupied in cold weather since they live in more protected areas. Spider sits in the center of the web; open hub. Spider may retreat to nearby cracks and crevices.

Observations: All of the webs of this spider I have found have been along the faces of cliffs; mostly under overhangs. Webs that I have seen tended to be more vertical than horizontal. Spiders sits in center, inverted.

Cave Orbweaver *Meta ovalis*

Spider

Body size is about 10 mm with a legspan of about twice this. The oval abdomen is yellowish to brown with a light pattern on it. Spiders may be seen on the webs from spring all the way to fall.

Thickjawed Orbweaver — *Family Tetragnathidae*

Web—Orb Web

Of all the orbweavers, this species is least likely to make webs. As adults, they do not make any catching snares, but hunt prey is a more active way; living under rocks, stones or dense vegetation. Webs may be made only by young. Web is horizontal; small to medium with about 20 radii. Webs can be seen in grassy areas before their final growing molt.

Observations: I have seen these spiders often and they seem to be fairly common; I have even seen them on the surface of the snow. Only once have I seen one, an immature, in a web. It was on a summer morning. The web was coated with dew and the spider was in the center of the open hub of this horizontal web. Web was just a few inches above the ground. A rare sight.

Thickjawed Orbweaver *Pachygnatha autumnalis*

Spider

Body length is 6 or 7 mm long with a legspan of 10 to 15 mm. The oval abdomen is brown with light and reddish pattern in the center. A light band runs along the side of the abdomen. Carapace is dark. The chelicerae are large and thick; this is why the spider is called the Thickjawed Orbweaver.

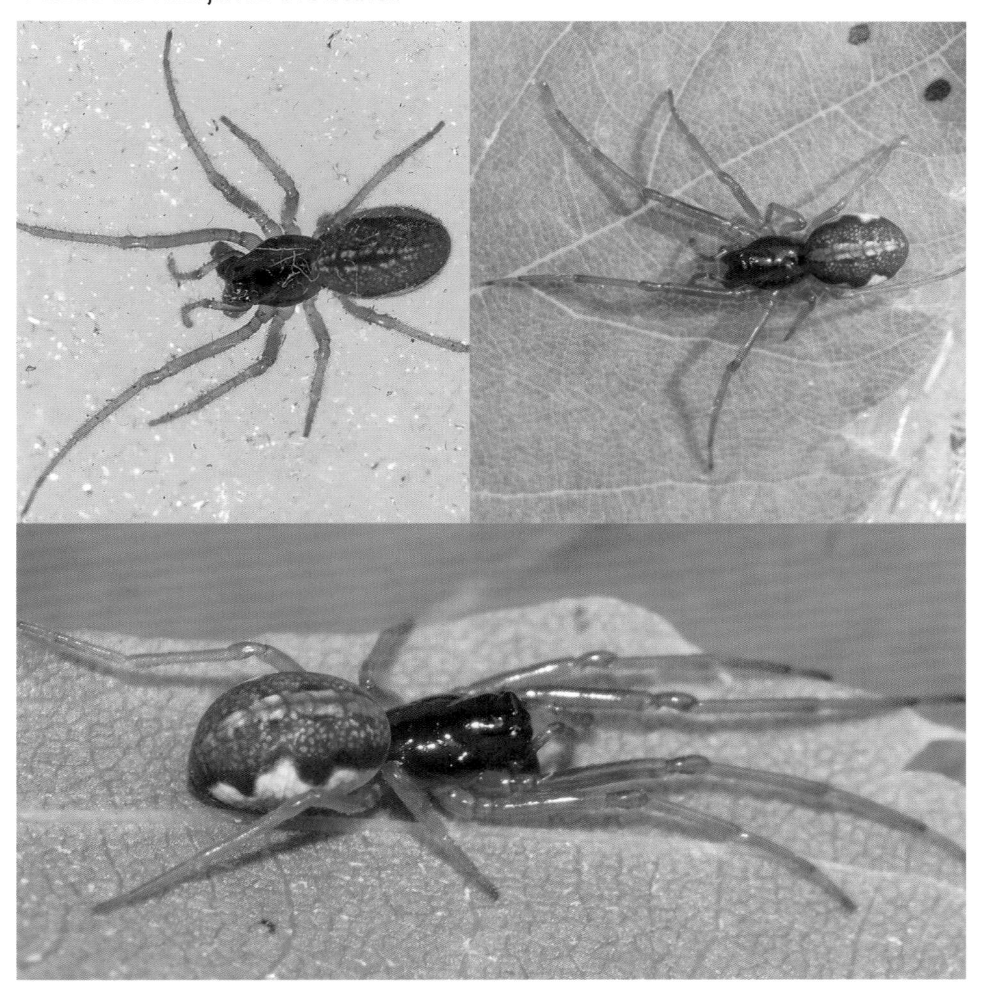

Longjawed Orbweavers | *Family Tetragnathidae*

Web—Orb Web

Webs of these spiders are very common in summer, late summer and early fall. The webs are mid-sized to large; up to 2 feet with 12 to 20 radii and an open hub. Webs may be vertical or horizontal with widely spaced spirals. The webs are most common in wetlands, but also found in meadows, yards and gardens. Webs are built at dusk and used at night; they are often taken down in the morning. Spiders do not make a retreat, but will hide along the edge. With a long thin body, they can stretch out and be hard to see.

Observations: It is hard to visit the edge of a lake or marsh on a late summer's evening and not see these webs. There are several species in the area, but they all make similar webs, along the wetlands. I have found their webs to be vertical or horizontal, but the latter seems to be more common. I have also found their webs in woods, but these appear to be made by smaller, younger spiders.

Longjawed Orbweavers | *Tetragnatha* species

Spider

Bodies are long and thin, 5 to 15 mm with long legs giving a legspan of 25 to 50 mm. The long thin abdomen may be light brown to nearly yellow. Very long chelicerae give the name of Longjawed Orbweavers. Very long legs have caused these spiders to also be called stilt spiders.

Tetragnatha Webs

Tetragnatha Webs

Triangle Web Spider — *Family Uloboridae*

Web—Orb Web; Triangular

These small triangular webs with only four radii are unique. At first glance they appear to be a partial orbweaver web. Spiders form these radii into a triangle shape. At the site where the radii come together, the spider extends a thread and holds it, becoming the anchor of the third corner of the triangle and holding the web to a tree or shrub. The spider actually becomes part of the web. Webs are found in late summer in deciduous trees or among understory shrubs.

Observations: As a tiny spider with a tiny web, it is easy to overlook this snare. But with the right angle of light, the web can be seen and subsequently the spider can be found. No other web is triangular and has the spider as part of the thread. I only recently and joyfully found my first web and spider in northern Minnesota.

Spider anchors web to branch

Triangle Web Spider *Hyptiotes cavatus*

Spider

Very small; 3 to 4 mm with a legspan of about twice this size. Abdomen is broadly elliptical and brownish gray; with a double row of cone-shaped bumps along the sides. Not likely to find spiders outside of the web.

This species normally hangs inverted in the web. The spider uses its own body to hold the third corner of the web to a tree or shrub. In this way it can instantly detect when prey hits the web, after which it pulls the web taught and then quickly releases it to ensnare the victim.

Orb Webs

Featherlegged Orbweaver *Family Uloboridae*

Web—Orb Web

Web is small only four to six inches in diameter with many radii; maybe as many as 40. Web is a few feet above the ground in shaded woods of summer. Though the web is an orb, it is horizontal and not sticky (hackled). Webs also have an open hub with a stabilimentum.

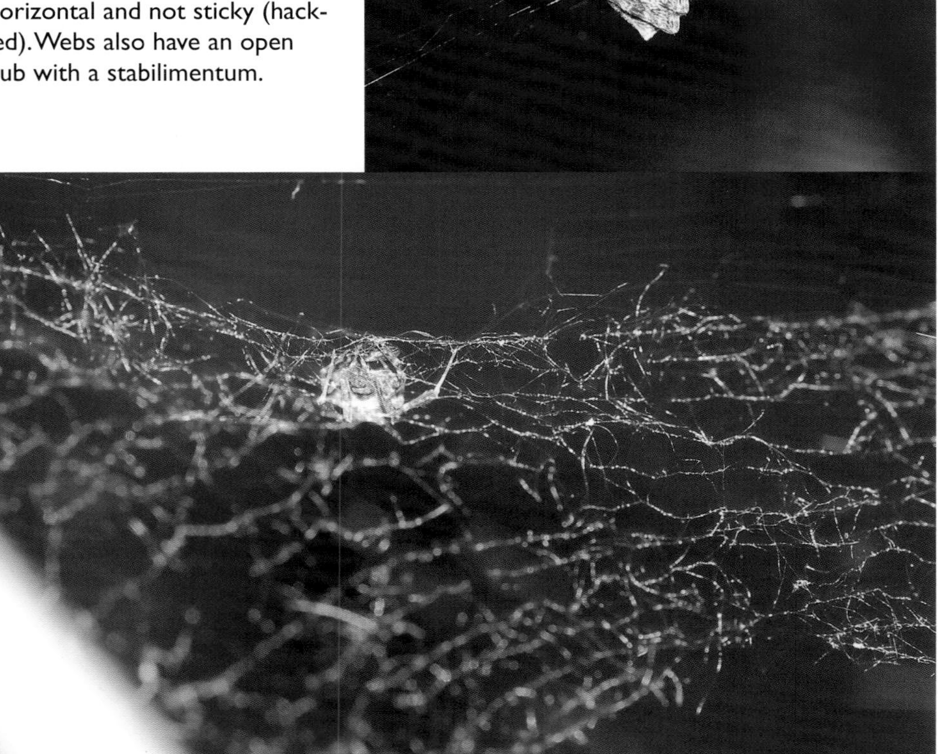

Observations: Though I have fairly often found these webs in summer (in shaded woods) I have never found many of them on a single day. With radii, stabilimentum and spirals, the web seems to be quite complicated.

Featherlegged Orbweaver *Uloborus glomosus*

Spider

Small body of 2 to 5 mm with a legspan of about twice this. Front legs are longer than the rest of the body and at the tibia, they have the "feathery" growth. Abdomen is triangular with a pair of dorsal humps. Abdomen is light brown. Spider hangs under the horizontal web.

Insects caught in Webs

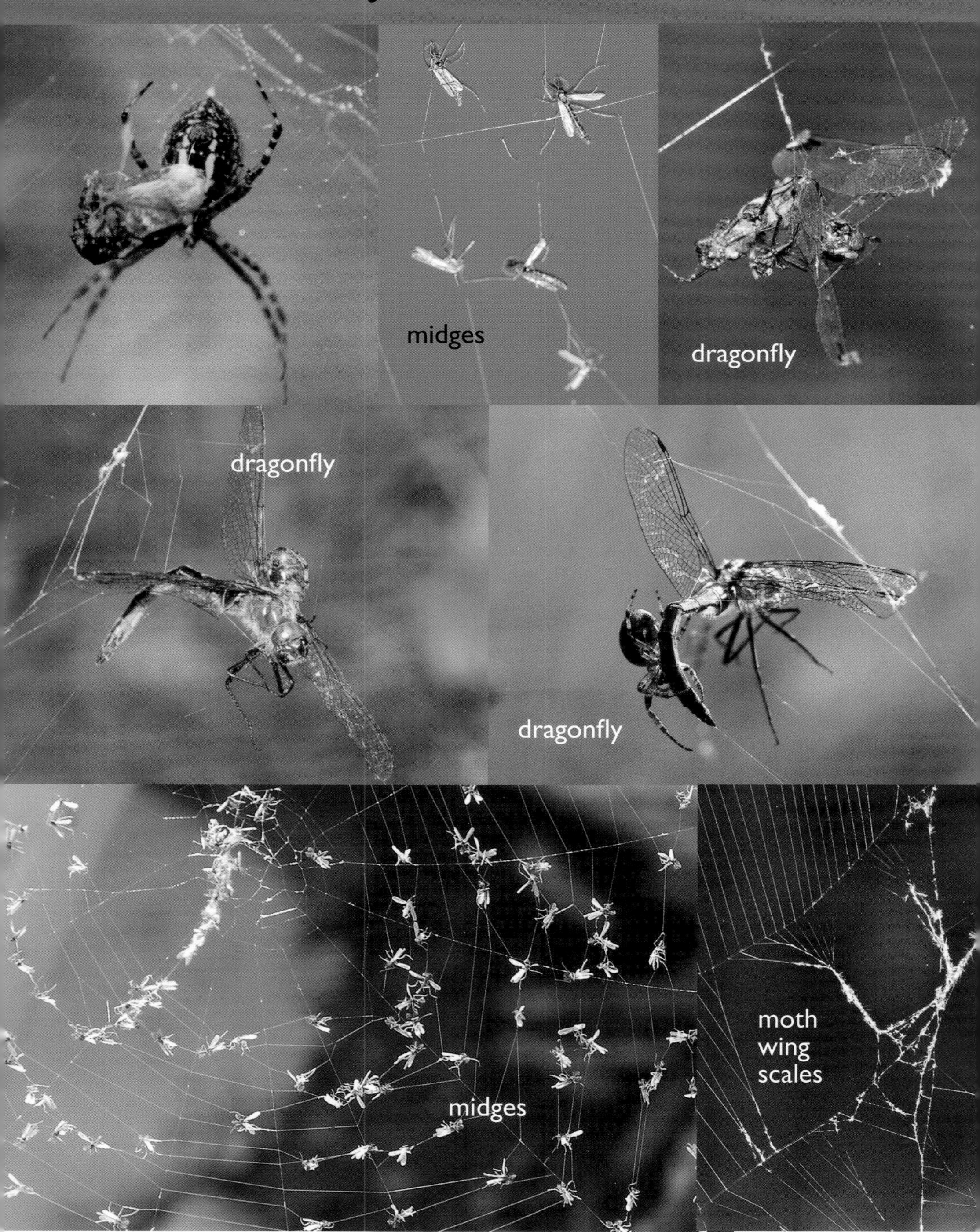

Insects caught in Webs

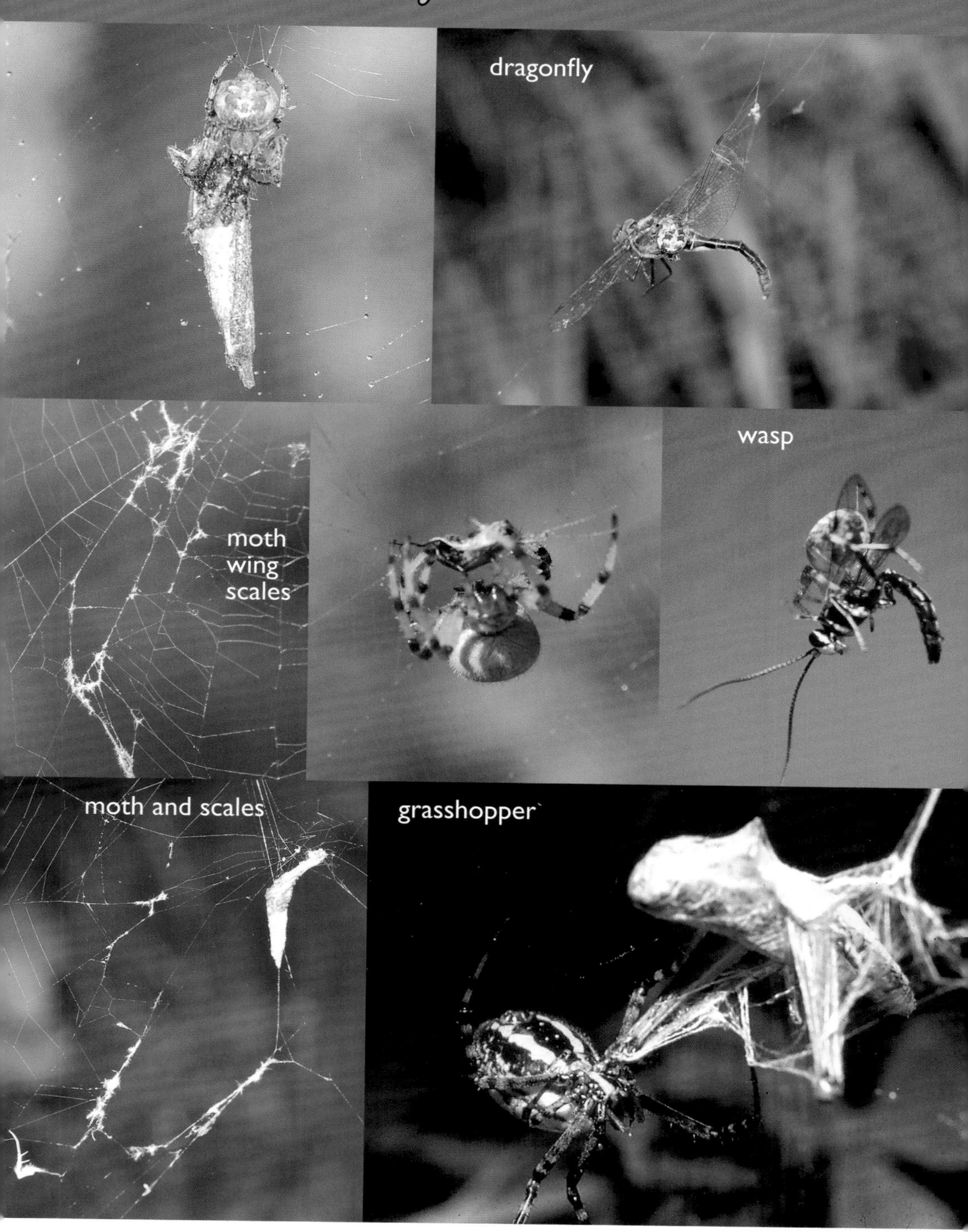

Dewy Webs

Dewy Webs

Frosty Webs

Snowy Web

A rare photo of a snow-covered web taken after a late season snowfall in May.

Dusty Webs

Seeds in Webs

Other Silk Constructions

Other Silk Constructions

Other Silk Constructions

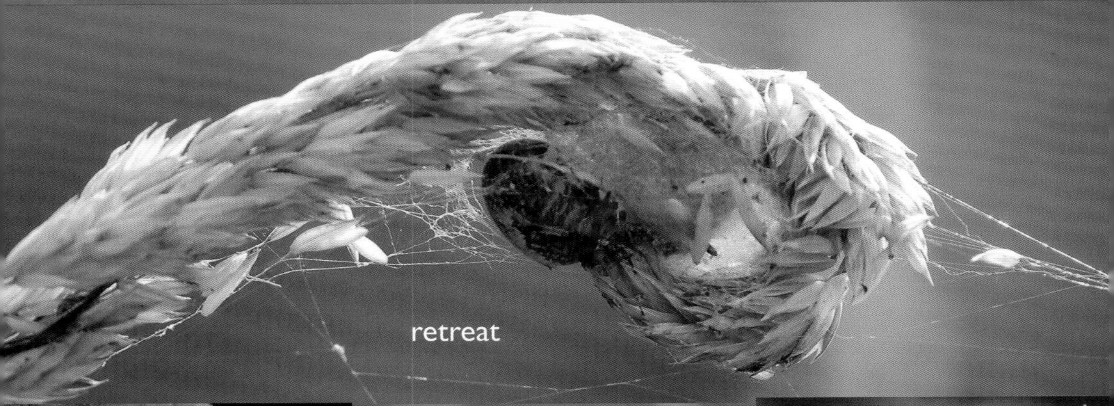

retreat

crab spider guarding her egg sac

retreat

retreat

nursery web

retreat

Other Silk Constructions

Man-made Web Substrates

Titles of Interest

Bradley, R.A. 2013. *Common Spiders of North America.* Berkeley, CA: University of California Press

Brunetta, L. and C. Craig. 2010. *Spider Silk*. New Haven, CT; Yale Univ. Press

Dondale, C. D. and J.H. Redner. 2003. *The Insects and Arachnids of Canada; Part 23*. Ottawa, ON, Canada: Department of Agriculture.

Foelix, R.F. 1996. *Biology of Spiders*. New York, NY: Oxford University Press

Gaddy, L.L. 2009. *Spiders of the Carolinas*. Duluth, MN: Kollath-Stensaas

Hillyard, P. 1994. *The Book of the Spider.* New York, NY: Avon Books

Howell, W.M. and R.L. Jenkins. 2004. *Spiders of the Eastern United States*; Boston, MA Pearson Education

Kaston, B.J. 1978. *How to Know the Spiders*. Dubuque, IA: W.C. Brown Co.

Kaston, B.J. 1981. *Spiders of Connecticut*. Bulletin 70. State Geological and Natural History Survey of Connecticut.

Levi, H.W. and L.R. Levi. 1996. *Spiders and Their Kin*. New York, NY: Golden Press

Shear, W. 1986. *Spiders; Webs, Behavior and Evolution.* Stanford, CA: Stanford Press.

Weber, L. 2013. *Spiders of the North Woods: 2nd Edition*. Duluth, MN. Kollath -Stensaas Publishing

Photo Credits

t=top b=bottom
r=right m=middle
l=left

PHOTOS

(All photos by the Author, except the following)

Parker Backstrom: 93 m, 105 l, 108 b, 109 tr

James Henry Emerton (1847-1931): 38 bl (from *The Common Spiders of the United States*. Ginn & Company. Boston. 1902)

James Emery: 2 tr (via Flickr.com with Creative Commons license)

Trace Fleeman y Garcia: 38 br (WikimediaCommons.org)

Kevin Friedt: 112 t (coveredinsevindust at English Wikipedia via WikimediaCommons.org)

Judy Gallagher (Flickr): 57 b, 97 b, 101 all, 103 tr tl, 107 tl b, 113 tl br, 115 tl tr, 125 b, 127 b (all via WikimediaCommons.org with Creative Commons license)

James Gathany: 31 br (Public Domain via Center for Disease Control)

Rich Hoyer: 103 b (via WikimediaCommons.org with Creative Commons license)

Brocken Inaglory: 30 b (via WikimediaCommons.org with Creative Commons license)

Tiffany Kersten/USFWS: 114 t (via WikimediaCommons.org with Creative Commons license)

Dave Marciniak: 133

Tom Murray (www.pbase.com/tmurray74): 51 all, 79 tr

Cassie Novak: 139 br

Tim Ross: 109 b (via WikimediaCommons.org with Creative Commons license)

Alejandro Santillana: 93 bl (Public Domain via Insects Unlocked, University of Texas at Austin)

Harvey Schmidt: 69 bl

Katja Schulz: 104 b (via WikimediaCommons.org with Creative Commons license)

Justin Sewell: 30 t (via Flickr with Creative Commons license)

Smidon33: 98 r (via WikimediaCommons.org with Creative Commons license)

Sparky Stensaas (www.ThePhotoNaturalist.com): 3 tl tm bl br, 7, 12, 16 tr ml, 35 tl b, 36 tl tr, 37 tl tr bl, 39 b, 41 tr b, 45 all, 47 b, 49 tr m, 55 br, 59 t, 69 br, 71 tr br, 73 br, 75 tl, 81 tl b, 83 tr br bl, 84 bm, 85 br, 87 tr, 88 tr bl, 89 tl b, 91 br, 92 all, 93 tl tr br, 95 br bl, 99 tr tl, 100, 102 all, 110 tr, 111 bl br, 112 bl br, 113 bl tr, 115 mr br bl, 119 tr b, 132 b, 136 b, 137 t, 139 ml

Konrad Summers: 31 bl (via WikimediaCommons.org via Flickr with Creative Commons license)

ILLUSTRATIONS

page 34: from *How to Know the Spiders*, Kaston 1953

Billy Anderson (www.BillyAndersonArt.com): 9 b, 10, 17, 18, 24 all (all based on previously published sources)

Rick Kollath (www.kollathdesign.com): 4 all, 5 all, 7, 8, 9 t, 13, 14, 15 all, 21 all, 28 all, 38 tr tl, 42, 52 all, 60 all, 124 br

Index

Index